Marcel Schmitt

Slot die coating of lithium-ion battery electrodes

Slot die coating of lithium-ion battery electrodes

by
Marcel Schmitt

KIT Scientific Publishing

Dissertation, Karlsruher Institut für Technologie (KIT)
Fakultät für Chemieingenieurwesen und Verfahrenstechnik, 2015

Tag der mündlichen Prüfung: 18. Dezember 2015
Referenten: Prof. Dr.-Ing. Dr. h. c. Wilhelm Schabel
Prof. Dr.-Ing. Hermann Nirschl

Impressum

KIT Scientific Publishing

Karlsruher Institut für Technologie (KIT)
KIT Scientific Publishing
Straße am Forum 2
D-76131 Karlsruhe

KIT Scientific Publishing is a registered trademark of Karlsruhe Institute of Technology. Reprint using the book cover is not allowed.

www.ksp.kit.edu

Print on Demand 2016

ISBN 978-3-7315-0477-1
DOI 10.5445/KSP/000051733

Slot die coating of lithium-ion battery electrodes

Zur Erlangung des akademischen Grades eines

DOKTORS DER INGENIEURWISSENSCHAFTEN (Dr.-Ing.)

der Fakultät für Chemieingenieurwesen und Verfahrenstechnik

des Karlsruher Instituts für Technologie (KIT)

genehmigte

Dissertation

von Dipl.-Ing. Marcel Schmitt

aus Osthofen

Referent:	Prof. Dr.-Ing. Dr. h. c. Wilhelm Schabel
Korreferent:	Prof. Dr.-Ing. Hermann Nirsch
Tag der mündlichen Prüfung:	18.12.2015

Kurzfassung

Um den Klimawandel aktiv zu bekämpfen, initiierte die deutsche Bundesregierung im Jahr 2000 das "Erneuerbare Energien Gesetz". In diesem Rahmen sollen bis zum Jahr 2050 größtenteils regionale, erneuerbaren Energiequellen 80% des Energiebedarfs decken. Aufgrund deren unsteter Verfügbarkeit und der projektierten Anzahl von einer Million Elektroautos auf deutschen Straßen bis 2020, wird eine wirtschaftliche, elektrische Speichertechnik benötigt. Sekundäre Lithium-Ionen-Batterien (LIB) besitzen mit ihrer hohen Energiedichte, einer hohen Zyklenfestigkeit und einem vernachlässigbaren Memory-Effekt die hierfür passenden Eigenschaften. Zum aktuellen Zeitpunkt erschweren jedoch deren verhältnismäßig hohe Investitionskosten eine weitverbreitete Anwendung, sowohl im stationären als auch im Mobilitätssektor. Günstigere Lithium-Ionen-Batterien können deshalb dabei helfen, den Übergang hin zu erneuerbaren Energiequellen zu beschleunigen.

Im Gegensatz zu den relativ hohen Rohstoffkosten, bietet vorallem die Batteriefertigung Einsparpotenziale. Innerhalb der kostenintensiven LIB-Prozesskette, ist einer der wichtigsten Schritte die Herstellung der Elektrodenfolien. Erhöhte Produktionsgeschwindigkeiten und ein geringerer Ausschuss könnten dazu beitragen, diese Kosten zu senken. Geschwindigkeitslimitierend in dieser Prozesskette ist das vordosierte Schlitzgußbeschichten. Hierbei formt das dispergierte Elektrodenmaterial zwischen Gießer und dem zu beschichtenden Substrat eine Flüssigkeitsbrücke. Obwohl dieses Auftragsverfahren bereits oft in der Literatur diskutiert wurde, sind die Auswirkungen der relativ hochviskosen Elektrodenpaste auf das Beschichtungsbild im Detail noch nicht hinreichend bekannt. Gegenstand dieser Arbeit ist deshalb die wissenschaftliche Ausarbeitung eines fundierten Prozessverständnisses folgender vier Themengebiete:

- Optimal genutzte Prozessfenster für hohe Bahngeschwindigkeiten,
- Grenzen simultan-mehrlagig beschichteter Elektrodenfilme für eine erhöhte Zellperformance,
- Minimierte Ränder zur Produktionsausschussreduzierung und
- Maximierte Geschwindigkeiten beim intermittierten Beschichten für einen effizienten Kalandrierschritt.

Aus wirtschaftlicher Sicht ist die interessanteste Betriebsgrenze die maximale Beschichtungsgeschwindigkeit bei einer vorgegebenen Schichtdicke. Dabei ist eine geschlossene, homogene Beschichtung nur innerhalb bestimmter Prozessgrenzen möglich. Obwohl die Fachliteratur diese vielfältig thematisiert, sind die Auswirkungen der hochviskosen, scherverdünnenden und partikelbeladenen Batteriepasten auf die Prozessgrenzen noch weitgehend unbekannt. Zur Validierung bestehender Modelle validieren und um Besonderheiten des rheologischen Verhaltens aufzuzeigen, wurden in dieser Arbeit experimentelle Beschichtungen mit LIB-Elektrodenpasten und einer newtonischen Modellflüssigkeit durchgeführt. Die Versuchsergebnisse zeigten, dass im relevanten Geschwindigkeitsbereich dünne Beschichtungen beider Flüssigkeitssysteme von einem frontseitigen Lufteintrag limitiert sind. Relativ dicke Filme konnten hingegen bis zu Geschichwindigkeiten von $u_W = 100\ m/min$ stabil beschichtet werden, was durch Modellberechnungen in dieser Arbeit bestätigt werden konnte.

Weiterhin birkt die Unterteilung eines Elektrodenfilms in mehrere Lagen diverse Leistungspotentiale für die spätere Zelle. Allerdings ist schon für Zweilagengießer zu erwarten, dass die veränderte Spaltregion die Prozessgrenzen stark beeinflusst. Um den Einfluss einer zusätzlichen Mittellippe auf das Strömungsfeld zu charakterisieren wurde deshalb in dieser Arbeit ein Zufuhrverhältnis in das bestätigte einlagige Prozessfenstermodell eingeführt. Das erweiterte Modell sowie experimentelle Ergebnisse zeigten, dass die minimal erzielbare Schichtdicke mit dem Zufuhrverhältnis abnimmt. Als Ursache verkleinerter Prozessfenster im Zweilagenauftrag konnte somit der zusätzliche Druckverlust im Beschichtungsspalt identifiziert werden. Ein Effekt welcher jedoch durch das Zufuhrverhältnis gesteuert werden kann.

Bei der Schlitzgußbeschichtung kann es an den Filmrändern zu Randüberhöhungen kommen. Bislang werden diese in der Literatur nur durch ein Modell für Filmextrusionen von Schmelzen beschrieben. In dieser Arbeit wurde deshalb dieses Modell speziell für die Anwendung auf die Schlitzgußbeschichtungen erweitert. Hierzu wurde ein Haftparameter λ in das Modell eingeführt, um die zuvor fehlende Substrathaftung eines beschichteten Films zu beschreiben. Zur Validierung wurden LIB-Elektrodenbeschichtungen bei unterschiedlichen Spaltverhältnissen durchgeführt. Hierbei zeigten die Ergebnisse eine stark ansteigende Randhöhe für größere Spaltverhältnisse. Für die gegebenen Bedingungen und von geringfügigen Abweichungen abgesehen, konnte dieser Trend sehr gut durch das neue Modell beschreiben werden.

Um die möglichen Linienkräfte im Kalandrierschritt zu erhöhen, werden LIB-Elektroden oft intermittierend beschichtet. Trotz zahlreicher Patente zum Thema fehlt bislang jedoch eine grundlegende wissenschaftliche Betrachtung des dynamischen Teils dieses Prozesses. Hierbei steht der Flüssigkeitsdruck im Gießer und dem Zuführsystem in direkter Relation zur aufgetragenen Nassfilmdicke. Für homogene Filme muss dieser Flüssigkeitsdruck nach jedem Beschichtungsstart möglichst schnell einen stationären Zustand erreichen. Um diese Nivellierung zu verkürzen, wurde in dieser Arbeit ein Berechnungsmodell entwickelt, um die stationären Systemdrücke des Beschichtens und Unterbrechens anzugleichen. Dessen experimentelle Validierung führte zu annähernd konstanten Systemdrücken bei vorhergesagter Filmdicke. Es wurden weiterhin die geschwindigkeitsbegrenzenden Mechanismen der intermittierten Auftragsverfahren diskutiert und festgestellt, dass die apparaturbedingten Totzeiten das Verfahren stärker limitieren als die relevanten Prozessfenster.

Abstract

To combat climate change, the German federal government passed the "Renewable Energy Sources Act" in 2000. For this purpose, mostly regional renewable energy sources are to cover 80% of all required energy, by the year 2050. Due to their dependence to ambient conditions and for a projected amount of one million electric cars on German streets by 2020, one still needs an economical electric storage technology. A reasonable approach is to use secondary lithium-ion batteries (LIB) which are characterized by a high energy density, large cycle stabilities and a negligible memory effect. To date, however, relatively high investment costs hinder their widespread use as a stationary or automotive energy storage technology. Thus, cheaper LIB technology could help to accelerate the transition towards renewable energy sources.

In contrast to the high raw material costs, the cell manufacturing processes usually offer several refinement approaches. In the costly LIB manufacturing chain, the processing of the electrode foils is one of the most crucial steps. Here, increased speeds in the production lines and reduced scrap rates could help decrease these costs. The rate determining step in this process chain is the pre-metered slot die coating. To produce a continuous film, the liquid has to bridge the small gap between the slot die and the web to form a stable coating bead. Although, this coating step is mentioned quite often in literature, the singularities of relatively viscous battery slurries and their effect on the film application were not scientifically discussed yet. The scope of this work is therefore the scientific elaboration of an enhanced process understanding of the four following topics:

- A maximal utilized process window in terms of high speeds,
- Limits of subdivided, simultaneous coated electrode films for an increased cell performance,
- Minimized edge effects to decrease waste-rates and
- High-speed intermittent coatings to enable for an efficient film calendering.

From an economic standpoint, the most interesting limit is the maximum coating speed for a given film thickness. Nevertheless, homogeneous coatings can only be obtained within certain coating limits. Although, these coating limits are already presented in literature, their impact on the highly viscous, shear-thinning and particle-loaded LIB coatings is still unknown. To validate the available coating window models for battery slurries, both, model predictions and experimental coatings were accompanied by those for a Newtonian liquid to reveal rheological singularities. The experimental results showed that highly viscous Newtonian coatings as well as LIB slurry coatings are limited by air entrainment in the relevant range of coating speeds. Affirmed by the model predictions, relatively thick coatings were shown to be stable even at $u_W = 100\ m/min$ without a film break-up.

Subdivided LIB electrodes facilitate several cell-performance improvements. However, even for bilayer slot dies, it is reasonable that the altered gap regime also highly affects the processing limits. Hence, to characterize the impact of the additional mid-lip pressure gradient, a top to bottom feed ratio was introduced into the before approved single layer model. The extended model confirmed the experimental results showing that the minimal achievable film thickness decreases with the adjusted feed ratio. Thus, the particular pressure profile of bilayer coatings could have been identified as the cause of the smaller bilayer coating process windows—a mechanism which can be controlled by the feed ratio.

Slot die coating further can result in the formation of unfavorable super-elevated edges(Dobroth and Erwin, 1986; Gutoff et al., 2006; Schmitt et al., 2012b; Schmitt et al., 2013d; Bitsch et al., 2014). To date, the literature only contains a model for super-elevations in hot melt film-extrusions. In this work, this model was adapted and extended for slot die coating. Since, the film's adhesion to a substrate is not present in extrusion coating an adhesion-parameter λ was therefore invented. For the model validation LIB electrode films were coated for different gap ratios. Here, the results showed increasing edge heights towards higher gap ratios. With some minor deviations, the novel model predicted the observed trend for the given conditions quite well.

To increase the possible line forces in the film calendering, lithium-ion battery electrodes are often coated intermittently. Although there are many patents for intermittent slot die coating, the dynamic part of this process has not yet been covered by commonly-available scientific research. In slot die coating there is a direct relationship between coated wet film thickness and liquid pressure distribution in the die and upstream feed system. For casting a homogeneous film, this liquid pressure has to reach a steady state with a compulsory leveling at each start. To shorten this ramping period, a pressure calculation model was developed to align the system pressures of the steady coating and interruption states. This resulted in an approximately constant system pressure and the predicted film thickness. In further considerations, the speed-limiting mechanisms of the intermittent coating methods were discussed. It was found that the enforced interruption method limits the projected coating more than the relevant process window.

Content

Vorwort

Als Grundlage dieser Arbeit dienen die Forschungsergebnisse, welche in meiner Zeit als wissenschaftlicher Angestellter und Doktorand, von Februar 2011 bis Juli 2015, am Institut für Thermische Verfahrenstechnik (TVT) im Bereich Thin Film Technology (TFT) am Karlsruher Institut für Technologie (KIT), entstanden.

Meinen Vorgesetzten Herrn Prof. Dr.-Ing. Dr. h. c. Wilhelm Schabel und Herrn Dr.-Ing. Philip Scharfer möchte ich für die interessante Aufgabenstellung, die zahlreichen fachlichen Diskussionen und die gewährten Freiheiten bei der Durchführung meiner Studien danken. Sie ermöglichten es mir, meine Ergebnisse mehrfach vor einem internationalen Publikum zu präsentieren.

Herrn Prof. Dr.-Ing. Hermann Nirschl danke ich für die Übernahme der Zweitkorrektur dieser Arbeit.

Herrn Prof. Dr.-Ing. Matthias Kind und Herrn Prof. Dr.-Ing. Thomas Wetzel möchte ich für die stets freundliche Unterstützung von Seiten der Institutsleitung danken. Bei Louisa Kollmar, Caroline Klotz und besonders bei Margit Morvay, möchte ich mich für die immer hilfsbereite, administrative Unterstützung aus dem TFT Sekretariat bedanken. Für die Umsetzung und die Verbesserung zahlreicher Konstruktionen, welche diese Arbeit erst möglich gemacht haben, möchte ich der Belegschaft der TVT-Werkstatt Michael Wachter, Steffen Haury, Markus Keller, Andreas Roth, Stefan Knecht und Stefan Fink, danken.

Meinen Kollegen in der TFT-Arbeitsgruppe, Michael Baunach, Felix Buss, David Siebel, Susanna Baesch, Stefan Jaiser, Anna-Lena Riegel, Sebastian Raupp, Ralf Diehm und Tobias Fritzensmeier, danke ich für die stets freundlich-kollegiale und diskussionsfreudige Atmosphäre. Den ehemaligen Kollegen Joachim Erz, Benjamin Schmidt-Hansberg, Sibylle Kachel,

Katharina Peters und Lukas Wengeler danke ich für zahlreiche Vorarbeiten und ihrer Unterstützung während der gemeinsamen Zeit am Institut.

Viele meiner Ergebnisse wären ohne die Zuarbeit der studentischen Arbeiten von Pascal Junges, Robert Schneider, Miriam Vogt, Cornelia Bürkin, Dennis Wagner, Niklas Russner, Carolin Schober, Michael Wittek, Ulli Hammann, Laura Rolinger, Paul Kitz, Susanne Unger, Orhan Keskin, Marco Branghofer, Moritz Weiß, Julian Klemens, Ralf Diehm, Sandro Spiegel und Manuel Schuster nicht möglich gewesen. Es war für mich sowohl eine große persönliche, als auch wissenschaftliche Bereicherung, einen Teil ihres Werdegangs zu begleiten.

Herrn Prof. Marcio Carvalho danke ich für sein wertvolles fachliches Feedback in gemeinsam durchgeführten Untersuchungen.

Den Kollegen der Mechanischen Verfahrenstechnik (KIT-MVM) und den Mitarbeitern des KIT Projektes Competence E (KIT-PCE) danke ich für die interessante und produktive Zusammenarbeit.

Ich danke meiner Familie, insbesondere meinen Eltern, meinen Großeltern und meinem Bruder, welche meine Arbeit stets durch Ihr Interesse und Ihre Anregungen unterstützt haben. Vor allem aber danke ich meiner Freundin für ihre Unterstützung während der schwierigeren Phasen dieser Arbeit, ihr Verständis für flexible Arbeitszeiten und ihre Liebe.

Abschließend möchte ich dem Bundesministerium für Wissenschaft und Forschung (BMBF) und dem Zentralen Innovationsprogramm Mittelstand (ZIM) des Bundesministeriums für Wirtschaft und Energie, für die Finanzierung dieser Arbeit danken.

Karlsruhe, Juli 2015

Marcel Schmitt

Preface

This thesis is based on the scientific results which evolved out of my employment as a research assistant and PhD-candidate at the Institute of Thermal Process Engineering (TVT) in the group Thin Film Technology (TFT) at the Karlsruhe Institute of Technology (KIT), from February 2011 until July 2015.

I gratefully thank my supervisors Prof. Dr.-Ing. Dr. h. c. Wilhelm Schabel and Dr.-Ing. Philip Scharfer for providing me this interesting research topic, their steering to keep the course towards this manuscript, the professional discussions and the scientific freedom I enjoyed during my studies. With their support, I was able to present the following results to an international auditorium, multiple times.

I would like to thank Prof. Dr.-Ing. Hermann Nirschl for providing the second review of this thesis.

For their throughout pleasant support, I would like to thank the head of the TVT institute, Prof. Dr.-Ing. Matthias Kind and Prof. Dr.-Ing. Thomas Wetzel. I thank Louisa Kollmar, Caroline Klotz and especially Margit Morvay, from the TFT head office for their helpful administrative support. Without the manufacture and improvement of countless constructions by the TVT workshop-team, namely Michael Wachter, Steffen Haury, Markus Keller, Andreas Roth, Stefan Knecht and Stefan Fink, this work would be nothing but dusty theory.

My TFT-working group collegues Michael Baunach, Felix Buss, David Siebel, Susanna Baesch, Stefan Jaiser, Anna-Lena Riegel, Sebastian Raupp, Ralf Diehm and Tobias Fritzensmeier, consistently provided a friendly and cooperative atmosphere keen to debate. I thank our former collegues Joachim Erz, Benjamin Schmidt-Hansberg, Sibylle Kachel, Katharina Peters and Lukas Wengeler for all their preliminary work and their support during our common time at the institute.

The results presented in this work would not exist without the studental support of Pascal Junges, Robert Schneider, Miriam Vogt, Cornelia Bürkin, Dennis Wagner, Niklas Russner, Carolin Schober, Michael Wittek, Ulli Hammann, Laura Rolinger, Paul Kitz, Susanne Unger, Orhan Keskin, Marco Branghofer, Moritz Weiß, Julian Klemens, Ralf Diehm, Sandro Spiegel and Manuel Schuster. To accompany a part of their career was not only a scientific but also a personal enrichment to me.

I thank Prof. Marcio Carvalho for his valuable and helpful feedback during our cooperative investigations.

I am further thankful for the yielding teamwork provided by the collegues of the Institute of Mechanical Process Engineering and Mechanics (MVM) and the project Competence E (PCE) on the field of lithium-ion batteries.

I would like to thank my family, especially my parents, grand-parents and my brother for always supporting my work by an open interest and several incitements. But primarily, I want to thank my girlfriend for her support during the more difficult stages of this work, her generous understanding of flexible working hours and her love.

Finally I want to acknowledge the financial support of this work by the Federal Ministry of Education and Research (BMBF) and the Central Innovation Program for small and medium-sized enterprises (ZIM), provided by the Federal Ministry of Economic Affairs and Energy (BMWi).

Karlsruhe, July 2015

Marcel Schmitt

List of symbols

Abbreviations and indices

AE	Air entrainment
B	Bottom
C	Coating
Ca	Capillary number
Crit	Critical
D	Downstream
EV	Electric vehicle
F	Film
FEPS	Filtration, de-aerator and pumping system
FH	Feed hole
G	Gap
Grav	Gravimetric
I	Interruption
KIT	Karlsruhe Institute of Technology
LCO	Lithium-cobalt oxide
LF	Low flow
LFP	Lithium iron phosphate
Li	Lithium
LIB	Lithium-ion battery
LMO	Lithium-mangan oxide
LNO	Lithium-nickel oxide
M	Mid-lip
Max	Maximum
Min	Minimum
NiCd	Nickel-cadmium
NiMH	Nickel-metal hydride
PCE	Project Competence E
P	Pipe
Pb	Lead
S	Slot

SEM	Scanning electron microscope
T	Top
TFT	Thin Film Technology
TVT	Institute of Thermal Process Engineering
U	Upstream
VCM	Visco-capillary model
Vol	Volumetric
W	Web

Latin letters

Lower case		
d	Diameter	mm
h	Height (z-axis)	µm
l	Length (x-axis)	mm
p	Pressure	mbar
q	Specific flow rate	mm^2/min
r	Radius	µm
u	Speed	m/min
v	Volume	ml
w	Width (y-axis)	mm
x	x-Axis	mm
y	y-Axis	mm
z	z-Axis	µm

Upper case		
E	Gap to height ratio	-
G	Gap ratio	-
M	Feed ratio top to bottom layer (3.26)	-
N	Factor	-
W	Neck-in ratio	-

Greek letters

$\dot{\gamma}$	Shear rate	1/s
ε	Power law exponent	-
κ	Consistence factor	Pa s^n
η	Viscosity	Pa s
τ	Shear stress	Pa
ϵ	Elongation ratio	-
λ	Adhesion parameter	-
Θ	Static contact angle	°
ϑ	Dynamic contact angle	°
Φ	Physical variable	

1 Introduction

1.1 Motivation

In 2000, the German federal government passed the "Renewable Energy Sources Act" to combat climate change (Deutscher Bundestag, 2000). Here, the transition towards renewable energy sources was addressed by determining their fixed feed-in remuneration (Deutscher Bundestag, 2004, 2008, 2011, 2012, 2014).

To accelerate this enormous project, one important goal was defined for future energy production. In 2050, mostly regional renewable energy sources are to cover 80% of all requested energy (German Federal Government, 2010). Even though technologies such as photovoltaics and wind power are adequately efficient (Hau, 2008; Konrad, 2013), they depend strongly on ambient conditions and are not always available. Thus, one still needs an economical electric storage technology with a high level of power availability and cycle stability (Gulbinska, 2014; International Energy Agency, 2014). A second goal to minimize carbon emission is to have one million electric cars on German streets and autobahns by 2020 (German Federal Government, 2009). To make this feasible, one requires not only affordable and highly available electricity, but also a high-density storage technology (Ketterer B., U. Karl, D. Möst, S. Ulrich, 2009; Ozawa, 2009; Thackeray et al., 2012).

A reasonable approach is to use secondary batteries to store the electric energy provided and consumed by such transient sources and consumers. Since they are characterized by large cycle stabilities and a negligible memory effect, lithium-ion batteries (LIB) might serve as a capable storage technology (Huggins, 2009; Ketterer B., U. Karl, D. Möst, S. Ulrich, 2009; Nazri and Pistoia, 2009; Ozawa, 2009; Thackeray et al., 2012; Yuan et al., 2012; Schlögl, 2013).

The reversibility of the LIB charging and discharging process results from a repeated intercalation and disintercalation of Li-ions into the active materials. To enable for an efficient ion-diffusion, the applied active materials and additives are arranged as thin-layered, particular films (Nazri and Pistoia, 2009; Yuan et al., 2012). In Figure 1.1, the operating process of a lithium-ion battery is drawn schematically.

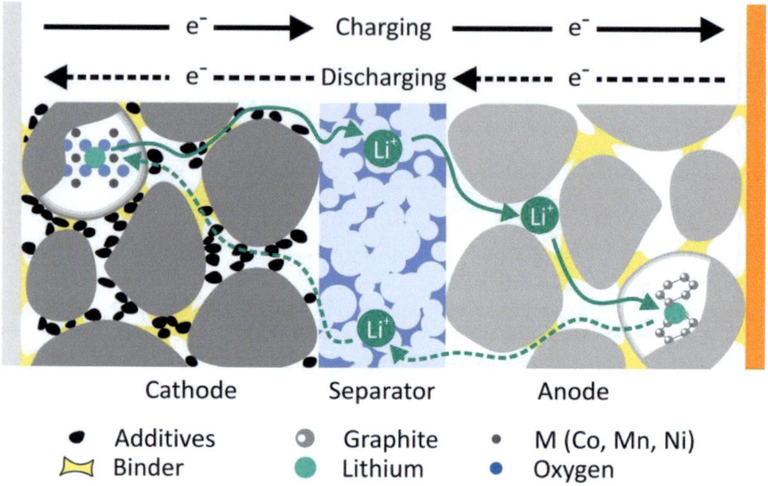

Figure 1.1: Schematic drawing of the components and operation of a lithium-ion battery cell.

To date, however, relatively high investment costs hinder the widespread use of lithium-ion batteries as stationary or automotive energy storage technology. Thus, cheaper LIB technology could help to accelerate the transition towards renewable energy sources.

The composition of investment costs for lithium-ion battery systems is very diverse. This is a result of the relatively long manufacturing and value chain from raw materials to a battery pack (Daniel, 2008, Consultants, Roland Berger Strategy, 2011a, 2011b; Neumann, 2011; McKinsey &

Company, 2012; Fraunhofer-Institut für System- und Innovationsforschung ISI, 2013; Kaiser et al., 2014) (As to see in Appendix 8.3). In Figure 1.2, the LIB manufacturing costs are graphically segmented and dedicated to the specific cost factors.

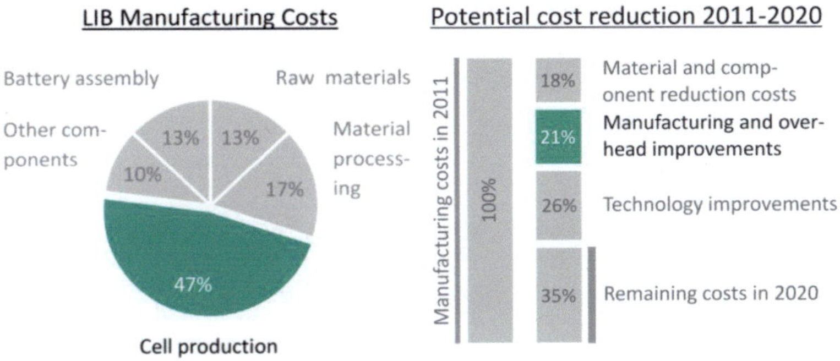

Figure 1.2: 47% of the lithium-ion battery investment costs are dedicated to the cells manufacturing (Consultants, Roland Berger Strategy, 2011b)(left). Nevertheless, studies consider the manufacturing step to have a high potential for cost reduction in the current decade (McKinsey & Company, 2012)(right).

According to several studies on the LIB cost structure, the cell manufacturing process is highly cost-intensive. In contrast to the also high raw material costs, however, manufacturing processes usually offer several refinement approaches. On the one hand, novel or revolutionary technologies can improve or replace some manufacturing steps (Huggins, 2009). These technologies may come from similar or completely different applications or even represent an as yet unknown invention. On the other hand, there is the economy of scale effect, which is capable of lowering specific production costs at nearly every manufacturing step.

For pushing both of these aspects further, the Karlsruhe Institute of Technology (KIT) launched the "Project Competence E" (PCE) in 2009. Connecting a multitude of international and interdisciplinary experts, the project allowed coverage of the entire value chain from raw material to battery module assembly. The thesis at hand was elaborated within the framework of PCE.

In the costly LIB cell manufacturing chain, the roll-to-roll processing of electrodes is one of the most crucial steps (Daniel, 2008; Neumann, 2011). Here, increased speeds in the production lines and reduced scrap rates could help decrease these costs. Figure 1.3 shows schematically the main process steps for stacked-cell electrode production.

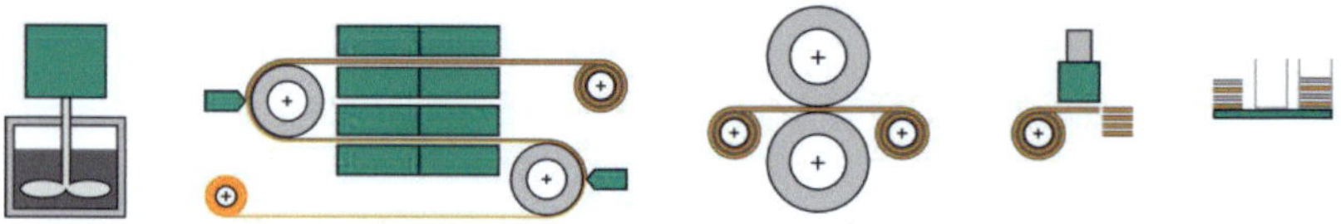

Figure 1.3: Schematic drawing of a typical process chain for stacked LIB electrodes (Schmitt et al., 2015a).

The typical LIB electrode film consists of active material particles, binder, and additives (Daniel, 2008; Nazri and Pistoia, 2009; Yoshio et al., 2009; Kaiser et al., 2014). The electrode has to be porous to allow for total penetration of the electrolyte as a lithium-ion transport path to the active materials. To deliver current, the electrode needs to be conductive and also connected to suitable metal foils, which serve as current collectors. For the film formation, the electrode components are mixed with a solvent and roll-to-roll coated on the metal substrate as a liquid film. Here, the most precise and common coating method is pre-metered slot die coating (Yoshio et al., 2009; Briec Emma, Müller Beate, 2015). In a subsequent drying step, the solvent is evaporated, leaving the desired porous film structure. To increase adhesion and energy density, the electrodes are calendered to a desired porosity (Nazri and Pistoia, 2009; Yoshio et al., 2009; Zheng et al., 2012). For a stacked cell layout, single stackable electrode sheets are then stamped out of the electrode rolls (Yoshio et al., 2009).

Due to a lack of scientific process knowledge, there are at least four efficiency draw-backs in the wet film processing of the electrode films. Their overcoming could help to either lower electrode costs or to enhance electrode performance.

First, the process limits for slot die coating of LIB electrodes is yet unknown. By utilizing the maximum possible production speed for example, the manufacturing costs of the electrodes could be reduced. However, the impact of the shear-thinning, highly viscous and particle loaded LIB slurries is not yet revealed.

Second, each cast electrode consists of only one layer having more or less homogeneously distributed ingredients (Daniel, 2008; Neumann, 2011; Kaiser et al., 2014). In contrast, an electrode made of several layers having different active materials could improve cell performance or capacity (Imachi et al., 2007; Chen et al., 2010; Tran, 2013). However, it is reasonable that the altered gap regime of a multilayer slot die also highly affects the processing limits, thus they cannot simply be adapted from single layer applications. However, this assumption has not yet been covered by scientific research.

Third, slot die coating of LIB-electrode films likely results in the formation of super-elevated edges (Dobroth and Erwin, 1986; Gutoff et al., 2006; Bitsch et al., 2014). Because this leads to problems in the downstream up-winding and assembly process, these edges must be trimmed off. Preventing these edge effects could save materials, eliminate a whole process step, or at least lower the specific process costs.

Fourth, LIB electrodes are often coated intermittently (Cholinski, 1993; Watanabe et al., 1996; Schmitkons et al., 1998; Iwashita et al., 1999; Sakai et al., 2002; Shinozaki and Sakai, 2003; Masuda and Watanabe, 2006; Ozawa, 2009; Janssen, 2011; Huth and Sekler, 2014). The available intermittent coating methods affect film homogeneity and are only feasible at low line speeds. A scientific understanding of this problem, which is still missing, might thus improve productivity and lower production costs.

Therefore, the scope of this work should be a close examination of these current limitations in LIB electrode coating technology and their solution approaches by the author. To better grasp the potential for improving slot die processes for today's lithium-ion batteries, it is reasonable to first understand their underlying physics. Thus, the following chapters will discuss in detail the state-of-the-art theory slot die coating.

1.2 State-of-the-art slot die coating

The key task of every coating process is to replace gas by liquid on a solid interface (Gates, 1999). From the liquid perspective in a roll-to-roll process, this expresses itself in the transformation of a pipe flow into a rectangular plug flow. The means of realizing this liquid casting and flow transformation are as varied as their fields of application. In general, all coating methods can be divided into self-metered and pre-metered processes. Self-metered coating methods, such as roll-, gravure-, dip-, comma bar- or blade coating, are very basic and robust processes. In general, coating with these methods is realized by a moving surface that entrains liquid out of a reservoir. On the one hand, this self-metering nature is beneficial, since it does not necessarily require a complex upstream metering system. On the other hand, it might be disadvantageous, when high precision coatings are required. For self-metered coating methods, the amount of deposited liquid not only depends on the coating parameters, but also on the liquid and environmental properties. If for example an increase in ambient temperature changes the liquid viscosity, the coated film thickness must be recalibrated. Extra labor costs and excessive wastage are the consequences (Cohen and Gutoff, 1992; Kistler and Schweizer, 1997).

Slot die and cascade coating processes are examples of pre-metered coating methods. Here, the coating tools need to be actively fed with liquid, where all of the metered liquid is cast onto the (to be coated) web. This permits active control of the coating thickness h via the coating width w, web speed u_W and volume flow $\dot{v}$ or the specific volume flow q:

$$h = \frac{\dot{v}}{w \cdot u_W} = \frac{q}{u_W} \tag{1.1}$$

The conception of these processes also allows for simultaneous multilayer casting of several films through multiple feeds. The disadvantage of pre-metered coating methods is their dependence on a precise metering of a pre-treated liquid, as described in appendix 8.3.2.

Both slot die and cascade coating can be applied in either bead or curtain mode. In curtain mode, the coating apparatus is arranged some tens of centimeters above the web. While the liquid leaves the apparatus as a curtain, it gets accelerated and drawn to a thin film by striking the moving web. This mode is capable of very high coating speeds. Nevertheless, to ensure curtain stability, low viscosities and minimum web speeds are needed (Kistler and Schweizer, 1997). Bead mode, in contrast, is realized by a very close arrangement of coating tool and web. To produce a continuously-closed film, the liquid bridges the small gap between coating tool and web and forms a stable coating bead. Two advantages in that mode are that the coating bead does not necessarily require large speeds to form thin films and that it allows for higher viscosities. In comparison to slot dies, cascades require lower-viscosity pourable liquids to flow down their inclined planes, where the liquids face the environment. Thus, slot die coating fulfills all the demands for producing LIB coatings (see appendix 8.3.2), which explains why it is the prevailing coating method in this field (Cohen and Gutoff, 1992; Kistler and Schweizer, 1997; Yoshio et al., 2009; Benhamou, 2011; Glawe, 2014; Mecklenburg, 2014; Briec Emma, Müller Beate, 2015).

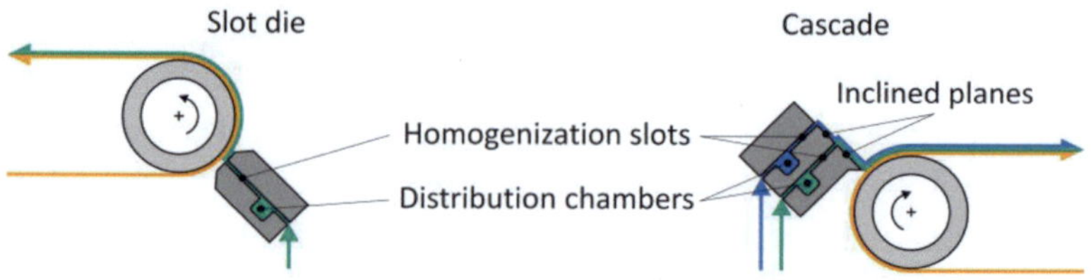

Figure 1.4: The layouts of a single layer slot die coater (left) and a bilayer cascade coater (right) in the bead mode.

As mentioned above, one key task of a coating tool is to distribute a liquid pipe flow homogeneously across the desired coating width. A slot die coater therefore has at least one distribution chamber with a neglectable pressure drop in the width direction. In contrast, there is a relatively high pressure drop in a following slot. Hence, the liquid flow is distributed inside the chamber preferentially in the width direction. The accuracy of this distribution is strongly related to the precision of the slot height and the pressure drop ratios (Kistler and Schweizer, 1997). Since the design requirements for precise liquid distribution are already well known (Sartor; Secor, 1997), they were not further investigated in this work. For simultaneous application of multiple layers, multilayer slot die coaters have several of these feed arrangements.

The second key task, film casting, takes place within the coating bead beneath the slot outlet lips. It has certain limits and singularities. Especially for high-viscosity LIB coatings, these are not yet fully understood. The following sub-chapters will discuss the state-of-the-art of limiting process windows, multilayer coatings, coating edge effects and intermittent coatings.

1.2.1 Process windows

The core requirement for almost every coating is a uniform and closed film. Indeed, the homogeneity of a coating is related to a stable coating bead, which can only be obtained within certain coating limits (Cohen and Gutoff, 1992; Kistler and Schweizer, 1997; Gutoff et al., 2006).

In slot die coating, the coating bead is located between the die lips - which follow the die slot - and the web. In web direction the coating bead is caged by a web-wetting upstream meniscus and a film-forming downstream meniscus. The bead stability is dependent on multiple parameters: the coating speed u_W, the gap height h_G, the resulting wet film height h, the up- and downstream pressure levels p_U and p_D, the die geometry, the liquid surface tension σ, and the viscosity η. A cross-section of the bead in web direction, which includes all relevant parameters, is shown in Figure 1.5. The 2-dimensional coating bead stability problem can be approximated as 1-dimensional within the gap. In this case, the lateral menisci are assumed to be of no significance with regard to continuous film process windows.

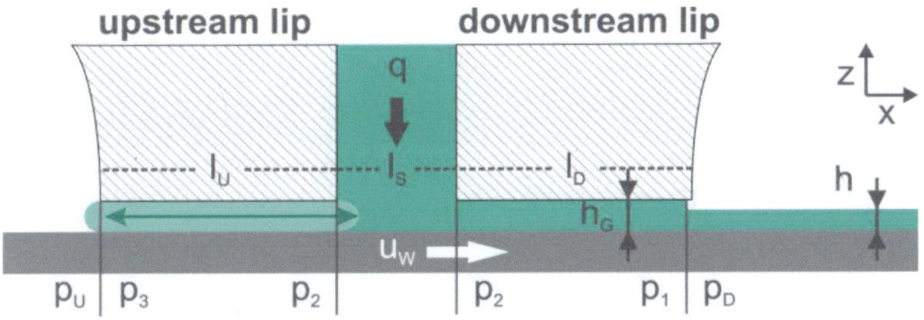

Figure 1.5: Schematic drawing of a slot die gap region showing all parameters relevant for bead stability.

On the one hand, coating bead stability depends on a consistent pressure balance between the environmental and bead conditions. If the pressure inside the bead is too high, liquid will swell out in all directions. Conversely, if the bead pressure is too low (e.g. for small films), air will be entrained upstream, thus breaking up the coating bead. This defect appears across the whole coating width simultaneously and starts with single air entrainment spots. If the pressure balance deteriorates, these spots connect to form uncoated streaks in cross-web direction.

On the other hand, bead stability also depends on an intact film-forming downstream meniscus. If the meniscus can no longer bridge the gap, beyond the so called low-flow limit, air breaks through from the downstream direction (Carvalho and Kheshgi, 2000; Romero et al., 2004). Since the meniscus breaks at certain points at the die lip, the characteristic film defects are uncoated streaks in web direction.

In the end the combination of applied process and liquid parameters will determine which of the break-up mechanisms (air entrainment or low-flow) is limiting the coating process.

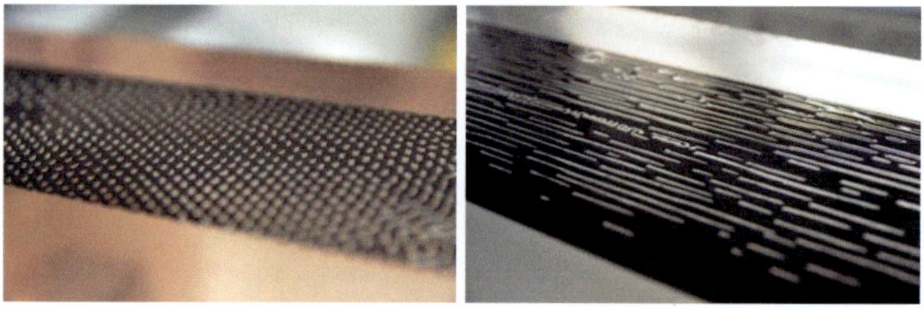

Figure 1.6: Film defects could appear as cross-web air entrainment (left) or web-directed low-flow streaks (right).

For economic reasons, the most interesting coating conditions are the maximum possible web speed for a given film thickness or conversely, a minimum film thickness for a given speed. Thus, the prediction of process windows was topic in many previous works to this coating method (Ruschak, 1976; Higgins and Scriven, 1980; Durst and Wagner, 1997; Carvalho and Kheshgi, 2000; Durst et al., 2007). Nevertheless, to date researchers have not investigated the impact of highly-viscous and particle-loaded LIB coatings on established process limits.

1.2.1.1 Air entrainment limit

The liquid flow in the upstream and downstream part of the coating gap induces a certain pressure drop. To predict the air entrainment limit, a pressure balance must therefore be applied to the wetted gap region. This balance has to equal 0 for ambient conditions around the slot die. In contrast, a vacuum chamber installed upstream to the slot die allows for a manipulation of the balance (Cohen and Gutoff, 1992). For Newtonian liquids the pressure drop equations for a coating gap flow are well known in literature (Durst and Wagner, 1997) and will be described more in detail in chapter 3.1.1. Depending on the location and shape of the upstream meniscus, four discrete cases of the gap pressure balance are possible (shown in Figure 1.7). The meniscus can be pinned at the front or at the feed-edge of the upstream lip. Further, it could be shaped concave or convex, which is displayed schematically in Figure 1.7.

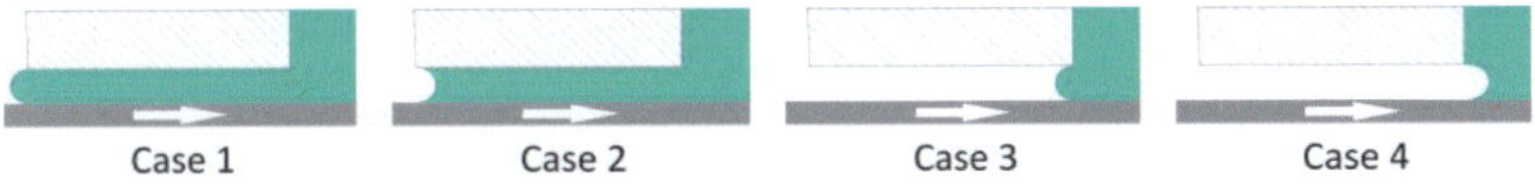

Figure 1.7: Schematic drawing of an upstream slot die section with the four stable pinning cases beneath the upstream lip. The white arrow marks the coating direction.

The cases 1 and 2 define limits for a stable located coating bead, where the upstream section is wetted and the upstream meniscus is pinned at the outer edge of the lip. If the resulting bead pressure gradient increases above those represented in case 1, the liquid gets pushed out at the upstream lip. If it decreases beneath those represented in case 2, the meniscus is no longer pinned at the upstream corner and starts moving towards the feed slot. For a pressure gradient in between the representation of case 2 and 3, the upstream meniscus has no fixed point to be pinned to. Here, the meniscus is susceptible to external disturbances, which might affect film uniformity. However, homogeneous coatings are still possible in this region, until the process is limited by the low-flow mechanism, which will be described below. In cases 3 and 4 the upstream meniscus is pinned to the upstream corner of the feed slot. If the pressure gradient decreases below those represented in case 4, the upstream meniscus can no longer bridge the gap and the coated film will be subject to defects. Thus, case 4 can be declared as the onset-limit of air entrainment.

The pressure gradients of the four discrete cases are plotted for water at a film thickness of $h = 40\ \mu m$, a gap height of $h_G = 128\ \mu m$ and various coating speeds in Figure 1.8. The applied equations (3.8)-(3.11) will be discussed in detail in chapter 1. Without the usage of a vacuum chamber the pressure gradient remains $\Delta p_{DU} = 0\ mbar$ and air entrainment will inevitably occure towards higher web speeds.

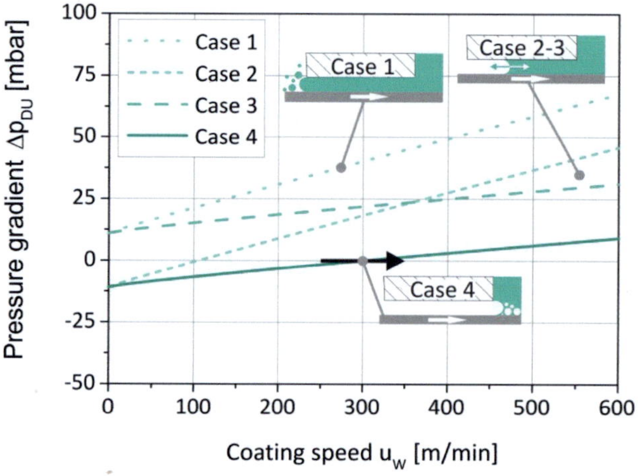

Figure 1.8: The process window, as calculated with equations (3.8)-(3.11) in chapter 1 for water and a wet film thickness of $h = 40\ \mu m$ at a gap of $h_G = 128\ \mu m$. The possible stable bead pressures are drawn for various coating speeds.

In industrial production, there are many coating liquids with non-Newtonian viscosity behavior, such as polymeric solutions or dispersions. Since LIB slurries are non-Newtonian liquids (what will be discussed in chapter 2.1, Figure 2.2), it is reasonable to expand the above process window to include their viscosity behavior. This behavior shows a monotonic shear-thinning character in the relevant range of shear rates and thus can be described using a power-law approach.

Tsuda (2010) and Lee (2011) already solved the known pressure equation in a slot die coating gap, using power-law viscosity models. They even found the processing windows to be enhanced towards higher speeds for shear-thinning viscosity liquids. Further, higher viscosities were found to be more stable against air entrainment.

An application of these models as processing window for LIB coatings could help to increase the current maximum line speeds.

1.2.1.2 Low-flow limit

Along with the air entrainment limit, the low-flow limit also defines a minimum film thickness or maximum speed in slot die coating. Here, above a certain speed, the downstream meniscus cannot bridge the distance between die lip and film anymore. In particular for thin films ($h \ll h_G$), this might be a reason for film defects even if the above air entrainment limit takes no effect.

The yet most common model for the description of the low-flow limit was developed by Carvalho (Carvalho and Kheshgi, 2000) and is related to the work of Ruschak (Ruschak, 1976) and Landau and Levich (Landau and Levich B., 1942). This so-called visco-capillary model (VCM) relates the break-up of the downstream meniscus to the coating parameters. To allow for a comprehensive comparison of various fluids and devices, experimental coating results within the VCM are typically presesented in dimensionless form (Carvalho and Kheshgi, 2000; Romero et al., 2004; Yang et al., 2004b; Chu et al., 2006; Romero et al., 2006; Chang et al., 2007; Lin et al., 2010; Tsuda, 2010; Bhamidipati et al., 2011; Jang and

Song, 2013; Wengeler, 2014). Here, the film thickness can be replaced by the dimensionless gap height G:

$$G = \frac{h_G}{h}, \tag{1.2}$$

while the web speed and the liquid properties are represented by the capillary number Ca. In Figure 1.9 the common VCM, which will be described in more detail in chapter 3.1.1, is plotted.

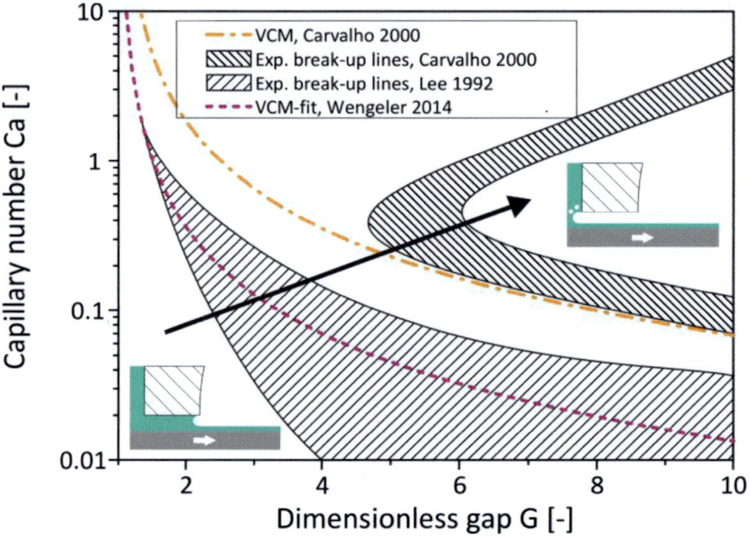

Figure 1.9: The low-flow limit can be described by the visco-capillary model (VCM). By plotting the critical capillary number for the dimensionles gap, the VCM is rendered dimensionless.

Since the VCM is not dependent on the bead pressure gradient, it appears at a constant limiting speed as a perpendicular line in the coating window of Figure 1.8.

Nevertheless, the VCM is based on a numerical derivation, valid for Newtonian liquids at low capillary numbers (Landau and Levich B., 1942).

Hence, the model might not be suitable for higher viscosities, coating speeds, or shear- thinning liquids. For liquids that are at least mildly viscous or particle-loaded, at capillary numbers $Ca < 1$, or for an un-pinned upstream meniscus, previous works showed a deviation from that theory (Lee et al., 1992; Romero et al., 2004; Yang et al., 2004b; Chu et al., 2006; Wengeler, 2014). Here, the onset of the low-flow limit is reported for thicker minimum film thicknesses or lower speeds. To fit the VCM to his own experimental results, Wengeler (2014) adapted the numerically-based model. The resulting graph is plotted beside the VCM in Figure 1.9. In contrast, other studies have also reported minimum film thicknesses below the prediction of the VCM for Ca close to unity and beyond (Carvalho and Kheshgi, 2000). Thus, beyond a certain capillary number, the low-flow limit once again allows for thinner films.

1.2.2 Multilayer coatings

Since many film-containing products (e.g. photographic, paper, surface protectional or battery applications) consist of multiple film layers, it would make economic sense to produce the film layers simultaneously (Nam and Carvalho, 2009). As a pre-metered process, slot die coating allows one to simultaneously cast multiple layers involving similar or different fluids (Cohen and Gutoff, 1992; Kistler and Schweizer, 1997).

There are several advantages for using subdivided layers when coating with lithium-ion battery slurries. One example, to pre-distribute additives and binder components by combining different slurries (see Figure 1.10; left). In doing so, the weight could be reduced and the cell performance improved. A second advantage is that different active materials could be combined in a single electrode (see Figure 1.10; right), which could also enhance the performance of the final battery.

At the same time, transport mechanisms during film-drying of polymeric components and sub-µm particles should be taken into account (Baunach et al., 2015). Experiments in this work showed a re-distribution of those components for multilayer LIB-coatings; this phenomenon is described elsewhere (see Appendix 8.2).

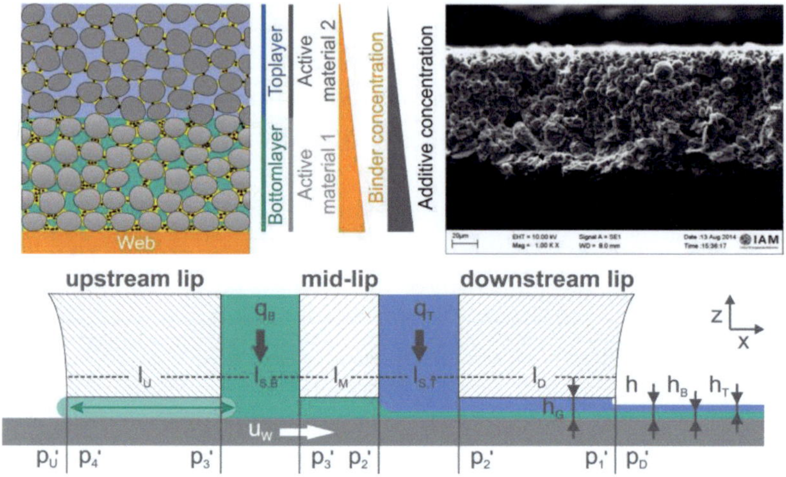

Figure 1.10: Schematic drawing of a two-layer LIB-electrode with pre-distributed active materials, binder, and additive components (top, left). SEM image of a bilayer, simultaneously-coated LIB anode with two different active materials between the top and bottom of the film (top, right). Schematic drawing of a bilayer, slot die gap region with all parameters relevant to bead stability (bottom) (Schmitt et al., 2015b).

Figure 1.10 *(bottom)* is a schematic sketch of a bilayer slot die gap with all relevant parameters. Here, in contrast to a single layer slot die, there is a second feed slot and a mid-lip. Inside the gap, the bottom and top layers form one common coating bead. Obviously, the resulting film thickness depends on the sum of the bottom and top feed flow rates:

$$h = h_B + h_T = \frac{q_B + q_T}{u_W} \tag{1.3}$$

The multilayer slot die coating differs from the single layer film application by virtue of several singularities. To attain homogenous multilayer films, these singularities must be considered.

In order to maintain the desired pre-distributed film morphologies, it is important not to mix the different layers during the coating step. Due to prevalent laminar conditions, convective mixing in the coating gap is unusual in slot die coating. However, in exceptional cases involving undesired bead conditions, vortices may indeed emerge in the gap (Kistler and Schweizer, 1997). Nam and Carvalho (2009) also discussed the phenomenon of a mid-gap invasion[1] of the separation point, which leads to fluctuation in the layer separation line. Here, the separation point is no longer pinned and starts moving from the top-layer feed slot in the upstream direction. This invasion occurs for a certain feed ratio and might provoke a vortex beneath the mid-lip.

Besides this, the multilayer coating of different liquids is only possible for a certain range of liquid properties. The solvents must be selected to avoid mixing and Marangoni-induced de-wetting. Here, diffusion-induced solvent mixing may increase the surface tension of the top layer (Horiuchi et al., 2015). Moreover, the viscosity ratio plays an important role in film stability. In general, smaller minimum films are reported for coatings with a lower bottom layer viscosity (Cohen and Gutoff, 1992; Yu, 1995; Lu et al., 2001). The ratio also affects the onset of mid-gap invasion and the amplification of external disturbances (Nam and Carvalho, 2010).

In summary, the single-layer process windows mentioned in chapter 1.2.1 cannot be applied to a bilayer process without modification (Cohen and Gutoff, 1992). All characteristics of the changed flow field must be taken into consideration. Available publications in the field of bilayer slot die coatings include numerical investigations and complex models to describe the flow inside the coating bead (Yu, 1995, 1995; Taylor, 1997; Hrymak

---

[1] *The separation line between top and bottom layer liquid (see Figure 1.10; bottom) moves upstream from its position at the lower end of the midlip.*

and Taylor, 1999). However, an impact of subdivided coatings on the limiting onset of air entrainment has not yet been observed and discussed. In this work, available break-up models will be adapted to compare the coating windows of single layer and bilayer applications in chapter 3.2. Partially these results are already published (Schmitt et al., 2015b).

1.2.3 Edge effects

Depending on the process parameters, slot die coating can induce super-elevated wet film edges - so-called heavy edges - at the side of the applied film (Dobroth and Erwin, 1986; Gutoff et al., 2006). These heavy edges must be avoided, since they lead to problems in the downstream winding process for example. Here, the edges of hundreds of wound layers undergo a superposition. A heavy edge of 10 µm in the dry film would superpose to a super-elevated edge of 1 mm by stacking 100 layers (or respectively in a wound-up coil with approximately 50 m of electrode). With cumulative amounts of several millimeters, these deformations cause undesirable, stiffening folds in the film, as shown in Figure 1.11. Furthermore, super-elevations cause inhomogeneous compression-force distributions during calendering, which lead to unfavorable current distributions in the operating cell. The effect of web-folding, shown in Figure 8.10, is additionally amplified by the presence of super-elevations. And finally, a stack of electrodes with heavy edges increases the overall cell volume, thus reducing the power density.

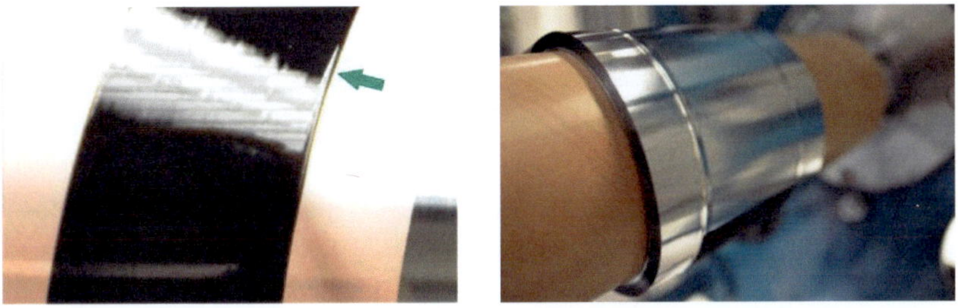

Figure 1.11: *Heavy edges in a wet LIB film (left) and their effect on a wound-up LIB electrode foil (right). The superposition of heavy edges leads to deformations in the mm-scale, which provoke a stiffening emboss in the metal foil and a delamination of the electrodes.*

Dobroth and Erwin (1986) listed three possible mechanisms as the source of super-elevated edges: die swell, surface-tension-driven edge formation, and edge-stress effects. The die swell effect cannot be the driving mechanism behind super-elevated edges in this case, since they are not only observed for swelling liquids. Moreover, LIB coatings are processed at high capillary numbers (> 1), which implies that surface tension forces have a minor impact. The focus here is placed on edge stress effects as a potential source of heavy edges. Dobroth and Erwin invented an expression to predict the height of super-elevations based on a neck-in of an extruded film. The origin of this neck-in is an acceleration of the average liquid velocity in the film direction.

Since the subject of Dobroth and Erwin's manuscript is hot melt extrusion, their model is only partially transferable and has to be adapted for slot die coating. In die extrusion, the hot melt transitions from a Poiseulle flow in the die slot to a plug flow outside the die. For slot die coating in contrast, there is an additional state in the gap between the die lips and the substrate. In this gap region, the liquid flow undergoes a superposition of a pressure-driven Poiseulle flow and a moving-wall-driven Couette flow. The crucial acceleration of the liquid happens between the flow state in the gap and the web-speed plug flow of the film. Thus, it should be affected by the chosen gap-to-film ratio.

Dobroth and Erwin's expression was compared to experimental results on LIB coatings (Schmitt et al., 2013c). Indeed, the results were not reproducible, due to non-steady-state measurements during start-up effects. Nevertheless, a general increasing trend for the width of the coating edges elevations was shown for increasing gap ratios.

Bitsch (Bitsch et al., 2014) showed that the coated edge height also depends on the low stress viscosity of LIB coatings. Higher zero-shear-rate viscosities seemed to hinder the formation of super-elevated edges. Thus, the formation may take place in the low shear stress transition zone before the film is fully resting on the coated surface.

So far, there is no scientific publication that contains a mathematical prediction of super-elevated edges in slot die coating. In this work a predictive model for the formation of heavy edges will be presented in chapter 1.

1.2.4 Intermittent coating

Intermittent slot die coating is becoming more and more important in a broad field of industrial applications (Kim S et al., 2009). For LIB-electrodes, this casting method even increases cell performance, due to higher possible line forces in the film calendering (see appendix 8.3.4). An intermittent slot die coated LIB film, manufactured in the KIT-TVT-TFT laboratory, is shown in Figure 1.12.

Figure 1.12: Intermittently coated LIB anode film on a copper foil on a pilot scale roll-to-roll coater, manufactured in the KIT–TVT–TFT laboratory.

In general, intermittent slot die coating consists of two more or less steady states—coating and interruption—separated by two hard to control transitions. At the coating start, the liquid and (depending on the process concept) parts of the machinery must be accelerated. As mentioned above (chapter 1.2.1), a stable coating bead between die lips and coated substrate is crucial for a homogeneous film. Therefore, the coating bead must be established as quickly as possible before the coating starts. In the second transition, the converse is true; all the liquid must be stopped and the coating bead removed quickly and without residues. Making things worse, LIB slurries are characterized by viscoelastic behavior. Consequently, slurry might leak out of the die slot during the interruption phase as it will be shown in the photographic image in Figure 5.1 in chapter 5.1.

Yang (2004a) initially described intermittent coating on a scientific basis. He found that the transitions between coating and interruption are comparable to leaving and entering the process window. In this regard, Chang (2009) discussed the start-up time of a coating bead and the parameters that influence it. Obviously, smaller coating beads are more quickly established. Kim (2009) further applied a bypass-valve for intermittent coatings and invented a measurement technique to investigate this method. In his work, he correlated the progression of liquid pressure in the slot die feed and cast film height. Despite all this previous work, an investigation of the comprehensive interactions between method, liquid pressure, and the resulting film is still missing.

There are mainly three different technology concepts available in industry for intermittent LIB slot die coating, as presented schematically in Figure 1.13. One concept includes a special bypass-valve to manipulate the feed (Watanabe et al., 1996; Iwashita et al., 1999), while a second concept uses slot dies with variable inner geometry (Schmitkons et al., 1998; Sakai et al., 2002; Shinozaki and Sakai, 2003; Masuda and Watanabe, 2006; Janssen, 2011; Huth and Sekler, 2014). A rapid build-up or break-up of the bead is often accomplished via a gap alternation as the third avialable concept (Szczepaniak et al., 2014; Vries, 2014). However, all of these technologies are optimized empirically on the targets of maintaining sharp edges and avoiding slurry leakage. So far none has been described theoretically. Moreover, due to mechanical movements, they are limited to lower web speeds.

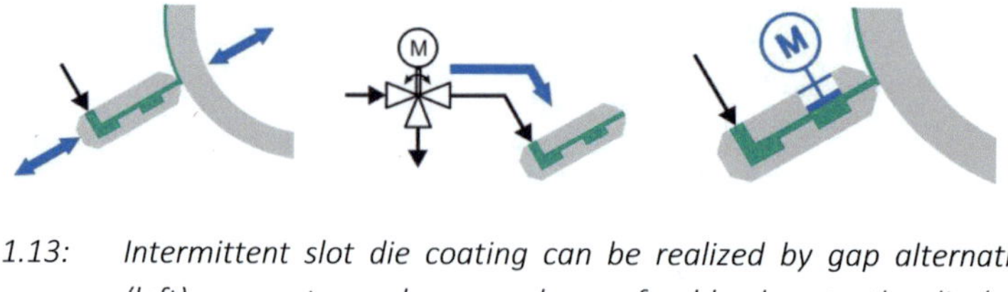

Figure 1.13: *Intermittent slot die coating can be realized by gap alternation (left), an upstream bypass-valve preferably close to the die (center), die-internal flexible volumes (right), or some combination of these techniques.*

In most cases, the latter mentioned intermitting concepts are combined with a bypass-valve. Thus, its effect on a process that is repeatedly leaving and entering a steady state should be investigated in more detail which is part of this work (see chapter 1) and already published (Schmitt et al., 2015a). A theorectical understanding could first reveal the dominating and limiting mechanisms of intermittent coatings and then help to control them.

1.3 Aim and scope of this work

The aim of this work is to gain a better understanding of the slot die coating process for the production of lithium-ion battery electrodes. In its scope it will therefore focus on the four following topics:

- The precise prediction of process windows
- The knowledge of these process windows also for bilayer coatings regarding subdivided, simultaneous coated films
- The forecast of super-elevated wet film edge formation
- The scientifical understanding of intermittent coatings towards high-speed film application

Coating limits such as air entrainment and low-flow are already presented in literature. Indeed, their impact on the viscous, shear-thinning and particle-loaded LIB coatings is still unknown. The hypothesis is that these models might as well be applicable to the slot die coating of LIB electrodes. For proving this assumption, the models will be characterized and compared to experimental observed LIB-coating process limits. An effect of complex rheological behavior of the battery slurries to the process limits should be observable by an experimental comparison to a Newtonian model system.

Subdivided LIB electrodes facilitate several performance improvements. However, multilayer slot die coating also includes certain characteristic singularities. It is reasonable that the altered gap regime highly affects the processing limits. However, this effect which has not yet been discussed should be assessable by adapting the available coating window models. To investigate this proposition, available single layer models will be extended to predict and discuss bilayer coating limits.

The mechanisms creating super-elevations at the coating edges remain particularly puzzling. To date, the literature only contains a model for super-elevations in hot melt extrusions. Due to its unpinned free liquid boundaries, however, extrusion coating is hardly comparable to coating onto a fixed surface. This work assumes, that an extension of this model to a slot die coating environment will solve this lack of knowledge. To verify of falsify the resulting edge formation model, it will be compared to experimental observed LIB coatings.

Intermittent slot die coating is yet a speed-limiting process. Literature showed a direct relation between liquid pressure in the coating feed system and the afterwards casted film. The hypothesis of this work is that one can reduce intermittent slot die coating to two steady states of coating and interruption and two preferably short transition states. On the one hand a pressure drop model for the feed system should then enable for a precise control and alignment of the steady states. On the other hand the duration of the transition states should be limited by the applied hardware set-up. For an experimental evidence of this hypothesis, the liquid pressure distribution in a feed system during intermitted slot die coating has to be characterized.

To investigate all these assumptions and to study the slot die coating process in detail, an experimental-set up, including slurry pre-treatment, several slot die geometries and a substrate-roller has to be constructed. It has to be capable of high-precision film measurements, even for high-speed coatings.

2 Experimental

In the following chapter, the liquid characterization methods, the experimental set-up and the experimental procedures used in this work will be presented.

2.1 Characterization of coating liquid

The knowledge of the rheology and the surface tension of the applied coating liquid is a key parameter for characterizing and predicting coating phenomena. The measurement of the applied liquids will be discussed in the following sub-sections.

The influence of the slot die coating process on LIB electrode films, was studied by the usage of water-based anode slurries (see Figure 2.1). These slurries were prepared by colleagues of the "Institute for Mechanical Process Engineering and Mechanics – Process Machines," within the common PCE project at KIT. The dry and wet mixing steps with an industrial-sized mixer (Inoue, Tri-Mix TX-15) will be described in appendix 8.3.1 in more detail. Due to continuous slurry improvements made by the partner-institute, the recipe and materials varied during the course of this work. Thus, the initial active material, a flaked graphite (Timcal, Timrex® SLP30, $d_{50} = 16\,\mu m$), was later replaced with a spherical graphite (Hitachi, SMG-A3, $d_{50} = 22\,\mu m$). A combination of waterborne latex dispersion (JSR, TRD 2001, styrene-butadiene rubber, SBR) and carboxymethyl cellulose (Daicell, CMC 2200, CMC) was used as a binder system. The cellulose also served as a thickening additive in the slurry (Kaiser et al., 2014). Additionally, a carbon black (Timcal, C-Nergy C65, CB) was added later on, which theoretically increases the inter-particular conductivity of an electrode film. The different anode recipes and materials used in this work are identified as slurries 1 and 2 in Table 2.1.

Table 2.1: Recipes in mass fractions of the applied LIB anode slurries.

	Slurry 1	Slurry 2
Flaked graphite	0.433	
Sperical graphite		0.536
CB		0.008
SBR	0.006	0.006
CMC	0.004	0.006
Water	0.556	0.444

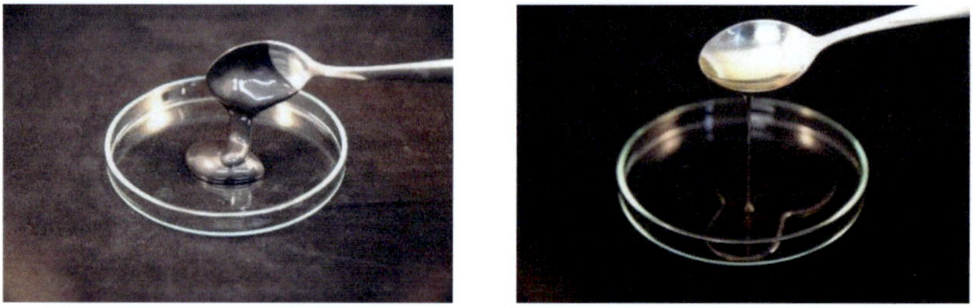

Figure 2.1: The highly viscous and shear-thinning anode slurry (left) and the silicone oil as a Newtonian model system (right).

Slurry 1 and 2 were highly viscous and showed a shear-thinning rheological character as it will be discussed in the next chapter. To illustrate the effects of this shear-thinning behavior, silicone oil (Carl Roth, Rotitherm® M220) as a Newtonian model system, was additionally applied as a comparison study to the coating trials.

2.1.1 Viscosity

For Newtonian liquids, the viscosity η acts as a proportional factor for the relationship between shear stress τ and shear rate $\dot{\gamma}$:

$$\tau = \eta \cdot \dot{\gamma} \tag{2.1}$$

The law of Oswald and De-Waele for non-Newtonian liquids defines the relationship between shear stress and rate as a power-law (Böhme, 2000):

$$\tau = \kappa \cdot \dot{\gamma}^{\varepsilon} \tag{2.2}$$

Here, κ is the consistence parameter and ε, the exponent. By the compination of equations (2.1) and (2.2) the viscosity for non-Newtonian liquids is given by the expression:

$$\eta = \kappa \cdot \dot{\gamma}^{\varepsilon-1} \tag{2.3}$$

The viscosity of LIB electrode slurries is affected by the relatively high particle loadings of often more than 50 mass% and the jelly-like binder-solvent combination. HIgher particle loadings in the slurry result in decreased costs for the solvent evaporation in the drying step. But, high particle loadings increase the viscosity and the pressure drop during slurry pumping. This might lead to problems with metering, leak resistance, or even processablitity. Based on this the typically applied slurry recipes are always a compromise between the processing costs and the slurry processability. In addition to its role in the final film, the binder increases the slurry viscosity and its stability by restricting particle sedimentation.

A rotational rheometer (Anton Paar, MCR 101) with a parallel plate system as the sensing adapter (Anton Paar, PP25-SN30563, $h_{Gap} = 1\ mm, d_{Plate} = 25\ mm$) was used to measure the slurry and oil viscosities. The rheometer measures the needed torque during a shear rate ramping and computes the resulting viscosity. The values obtained were averaged for at least three runs to reveal measuring inaccuracies. The viscosity data varied slightly from one batch to another and was often dependent on slurry ageing. Hence, each slurry was measured right before the associated coating trials. The related viscosity data is given along with the experimental runs in this work. Three exemplary viscosity measurements are plotted as functions of the applied shear rates in Figure 2.2.

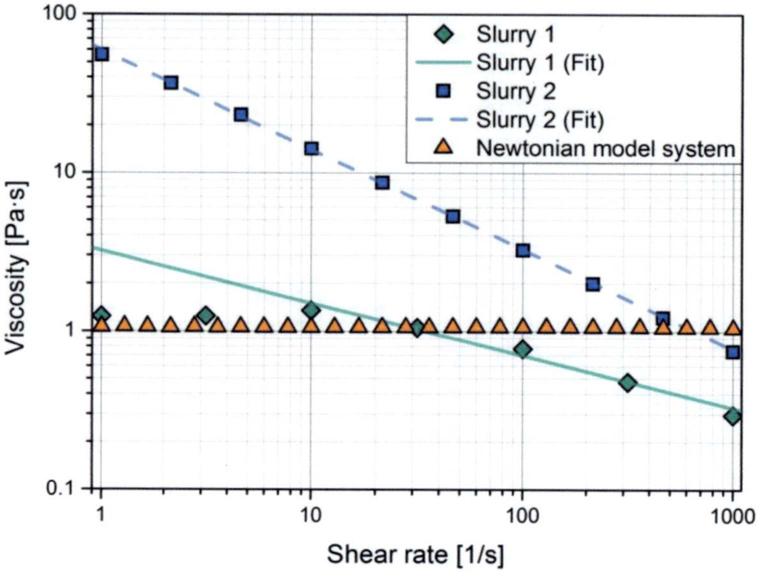

Figure 2.2: Viscosity measurements of the two slurry recipes and the Newtonian model system (silicon oil) for various shear rates. A power-law fit was applied to the shear thinning slurries for their prediction at $\dot{\gamma} \geq 10\ 1/s$.

The plot shows the shear-rate-independent viscosity of the Newtonian model system, which remains constant at $1\ Pa \cdot s$ in the whole range of applied shear rates. In contrast, the representative data of LIB slurries 1 and 2 exhibit shear-thinning behavior. Due to its lower solids content, slurry 1 has a lower viscosity, which is constant at around $1.4\ Pa \cdot s$ until $\dot{\gamma} > 10\ 1/s$. At higher shear rates, the viscosity decreases to $0.3\ Pa \cdot s$ for $\dot{\gamma} = 1000\ 1/s$. The viscosity of Slurry 2, with its higher solids content, starts at around $60\ Pa \cdot s$ for $\dot{\gamma} = 1\ 1/s$ and thins to $0.7\ Pa \cdot s$ towards $\dot{\gamma} = 1000\ 1/s$.

As mentioned in chapter 3.1, the power-law approach in equation (2.3) was applied to predict the shear-thinning viscosity. Thus, the parameters κ and ε were fitted to each viscosity measure. Slurry 1 was only applied to continuous coatings with shear rates above $10\ 1/s$ (see chapter 1) and the viscosity plateau below this rate was neglected for this fit. Shear rates above $1000\ 1/s$ are beyond the effective range of the available rotational rheometer. Yet, the decreasing trend in slurry viscosity was confirmed until $\dot{\gamma} \approx 100{,}000\ 1/s$ by means of a capillary rheometer showed by project partners in *Figure 4* in Kaiser et al. (2014).

2.1.2 Visco-elasticity

Newtonian liquids react to external stress with viscous flow. In contrast, the flow of visco-elastic liquids divides into a viscous and an elastic part (Mezger, 2010). Knowledge of the relationship between these parts is important for predicting two LIB slurry properties: sedimentational stability and response behavior. The latter might be important during intermittent coatings having a repetitive acceleration and retardation.

The rotational rheometer used for this study can measure visco-elastic behavior. While one plate is rotating with various deformations (amplitude sweep) or frequencies (frequency sweep), the rheometer is recording the resulting back-twist (Mezger, 2010). Thus, memory and loss moduli can be evaluated, which express the elastic and viscous behavior, respectively. A strong memory modulus indicates a jelly-like, and thus sedimentation-stable and elastic, behavior. Figure 2.3 shows a representative frequency sweep for slurry 2 and the Newtonian model system. Slurry 1 is not plotted in the below graph since it was only applied to continuous coatings.

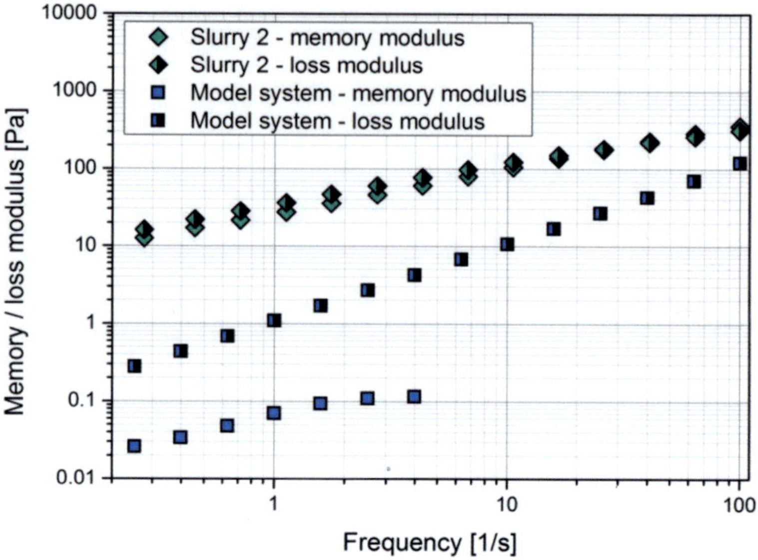

Figure 2.3: Frequency sweep of slurry 2 and the Newtonian model system, at deformations of 3.1% and 21%, respectively. In contrast to the Newtonian model system, the slurry shows a strong memory modulus.

The memory and loss moduli in the frequency sweep measurement of slurry 2 (deformation of 3.1%) are on approximately the same level. This implies that the slurry has a strong and non-negligible elastic part and might be related to entrained micro air bubbles during the mixing step. Although, the slurry was exposed to vacuum after the mixing and the deaerated in the pre-treatment (see chapter 2.2), its viscosity would prevent any movement of micro-scale sized air bubbles. In contrast, the subservient memory modulus of the Newtonian system is one decade smaller than its dominant loss modulus. This implies a more liquid behavior. Towards higher frequencies, the small memory modulus could not been detected by the rheometer.

In summary, the elasticity of slurry 2 should be taken into account for coating start-up times and, especially, intermittent coatings as dicussed in chapter 1. Effects based on this behavior should not occur when coating with the Newtonian model system.

2.1.3 Surface tension

The slurry surface tension was measured using the pendant drop method (Krüss, EasyDrop DSA20). This method is based on the comparison of an actual, measured drop-shape with a numerically calculated one (see appendix 8.5). If the two shapes are the same, the surface tension can be calculated on the basis of a known density difference (KRÜSS GmbH; Menges, 2002; Unger, 2013).

Toward this end, the LIB slurries were first filled into the instrument's syringe with a long cannula. By metering the liquid out of the cannula, a drop was formed at its tip. A microscopic picture of this drop was then analyzed by the instrument company's software (Krüss, "Drop Shape Analysis") using the above method. The highest reproducibility is achieved, if the drop repeatedly reaches its maximum size. Figure 2.4 shows a representative microscopic picture of a pendant drop of slurry 2.

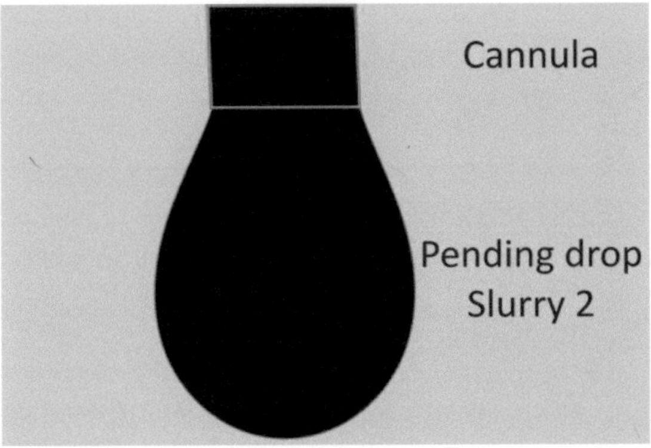

Figure 2.4: Representative microscopic picture of a pendant drop of slurry 2.

The measurement resulted in a surface tension $\sigma = 0.066\ N/m$ for slurry 2, compared to approximately $0.070\ N/m$ for pure water. Since varying the solid content produced only small deviations in this value, it was used for both water-born slurries. The surface tension of the Newtonian model system (silicone oil) was measured to be $\sigma = 0.018\ N/m$ (Unger, 2013).

2.2 Slurry pre-treatment

As described in chapter 1.2, the coating liquid pre-treatment is cruicial for homogeneous coatings. Film defects caused by air bubbles or agglomerates can falsify any results of the process investigations. Hence, a set-up with an acronym called "FEPS" (German acronym for filtration, de-aerator and pumping system) was created within this work, to pre-treat the delivered LIB slurries in batches. To ensure chemical stability, all parts coming into contact with the slurry were made of stainless steel.

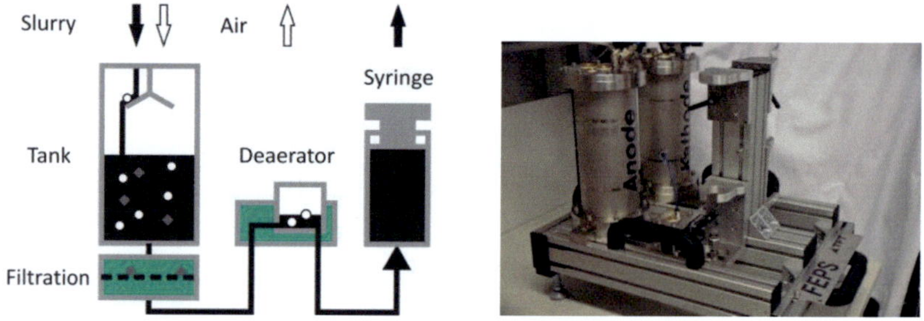

Figure 2.5: Schematic drawing of the individual sections of the FEPS (left). Photo of the FEPS design (duplicated to avoid cross-contamination between anode and catohode slurries) (right).

To prevent entrapped air pockets, a vacuum was applied during the filling of the 2 l FEPS tank. The slurry had to form a film while flowing over an inclined drainage, thus promoting the ascent of air bubbles. During the actual pre-treatment process, pressurized air pushes the slurry from the tank through a stainless steel filtration mesh. A mesh size of $250\ \mu m$ was found to be suitable for holding back most of the agglomerates. Subsequently, a de-aerator bled the slurry via an integrated air-scoop. Finally, a syringe was filled with the agglomerate-free and bled liquid, which was then ready to be metered for slot die coating.

Pumping the slurry through the pre-treatment with pressurized air has one major advantage: The whole system could then serve as a pressurized-tank metering-system with an inline pre-treatment. This was not necessary to be applied during the trials presented in this work, since only small slurry amounts were needed. A syringe pump was used instead, as described in the next chapter.

Usually the slurries were stored in the air-conditioned laboratory environment for a couple of houres before being processed. A temerature control step (as mentioned in appendix 8.3.2) was not necessary or required in this work.

2.3 Experimental set-up and procedure

The investigation of miscellaneous aspects of slot die coating requires several different experimental set-ups and settings. The basic set-up, its expansions, the applied slot dies and the experimental proceedings which were constructed within the scope of this work are presented and explained in detail in the following chapters and in more detail in the referred appendices.

2.3.1 Basic experimental set-up

The requirements on an experimental set-up to investigate process windows, edge effects, and intermittent methods in the slot die coating of LIB electrodes are quite diverse.

First, the equipment mounts must be very precise, so as to establish and maintain the relative positioning of die and coated surface. Sections 1.2.1-1.2.4 showed that the coating gap has a significant impact on the resulting wet film. Second, as the relative velocity between die and surface, the coating speed has to be adjustable over a wide range. Especially when characterizing process windows, the speed is an important limiting parameter. Thus, it should not be limited by the hardware. Third, one needs a means of evaluating the cast film and relating it to the influencing process parameters. To exclude any falsifications due to drying effects, the optimum would be to analyze only the wet film without drying effects. And fourth, a suitable experimental set-up must provide a certain amount of flexibility. It should be capable of mounting different kinds of slot dies or pumping systems.

To fulfill these requirements, a so-called "Development Coater" (TSE Troller AG) was used as the basis of the experimental set-up in this work. It provides a high precision stainless steel roller ($d_{Roller} = 350\ mm$) with an accurate slot die positioning system. The surface velocity of the roller, and thus the coating speed, is adjustable between 0.1-600 m/min (see schematically in Figure 2.6). In the course of several student theses, this basic apparatus was improved, so as to support various metering systems, self-constructed slot dies, and a film-analyzing tool (Junges, 2011; Schneider, 2012; Bürkin, 2012; Wagner, 2012; Vogt, 2012; Kitz, 2013; Diehm, 2014; Schuster, 2015). The roller bearings, however, were found to have a runout of approximately 20 µm in total. This had to be taken into account for each correlation of coating result and actual coating gap.

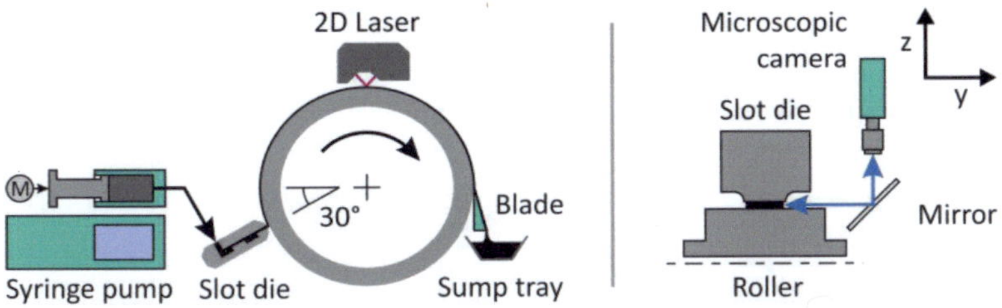

Figure 2.6: Schematic drawing of basic experimental set-up with syringe pump, slot die, roller and sensor system (left). Arrangement for recording the coating gap via microscopic camera (right).

The total amount of coated slurry needed for the trials in this work is rather small. Thus, a syringe pump was selected as the metering device (Chemyx, Nexus 6000). It can support a rack of four syringes at once. As applied in this work, syringe pumps offer two advantages over other mechanical pumping systems. First, they are not dependent on a performance curve, which facilitates volume flow control. And second, the slurry is already provided in syringes by the FEPS pre-treatment (see chapter 2.2).

The two-piece syringes used have a volume of $v_{Syringe} = 60\,ml$ and are made of a stiff Polypropylene/Polyethylene combination (HSW, Norm-Ject 60 ml). To increase the at-a-stretch usable slurry volume, the syringes were often combined to form a pack of two or four.

Since high liquid pressures were to be expected, the syringes and pipes feeding the slot die were bolted using a coupling ring/nut-system (Swagelok & Hi-lok). Due to their pressure resistance and form stability, compared to other flexible pipes, Teflon pipes were used ($d_{Inside} = 4\,mm$, $d_{Outside} = 6\,mm$). Moreover, Teflon pipes are more pliable than steel pipes, which offer certain advantages in terms of flexibility.

The various slot dies described in chapter 2.3.2 were placed at the 8 o'clock position relative to the roller. Since there was no vacuum chamber, the coatings were formed at ambient conditions. The gap height was adjusted at the thightest rotational position of the roller. In order to investigate wet-film stability, geometry, and quality, it is not necessary to coat the fluid onto a substrate and dry the film. In this work, the wet film was thus coated directly onto the high precision stainless steel roller. At the roller top, the wet film topography was then analyzed visually or by the mounted sensor system. After analysis, the film was scraped off the roller at its back.

To measure the film topography on-line, a 2-dimensional laser triangulation system was used (Keyence, LJ-V7060). This system has a relatively large detection width of 16 mm and an outstanding time-resolution of 16,000 Hz. The latter is important for analyzing the starting and stopping edges of high-speed intermittent coatings. During the coating trials, the sensor was arranged to always see some part of the uncoated roller. Thus, the coated film height was calculated directly from the measured step between film and the uncoated areas. To accomplish this, a prior inclinational correction of the sensor signal and the roller surface was necessary.

In addition to this height measurement, the experimental set-up provided the opportunity to view the coating gap (as shown in Figure 2.6). As mentioned in chapter 1.2.1.1, the location of the upstream meniscus can be an important indicator for coating defects. For this purpose, the microscopic camera of the pendant-drop measuring device was mounted temporarily beside the roller.

2.3.2 Slot dies

In this work, three self-designed slot dies with differing dimensions and capabilities were used.

All three slot dies have the same general construction, with two main plates separated by a U-shaped metal shim-foil. While the downstream plate provides the liquid feed-hole, the distribution cavities are machined in the upstream plate (as shown in Figure 2.7). The details of the three slot dies are listed in Table 2.2.

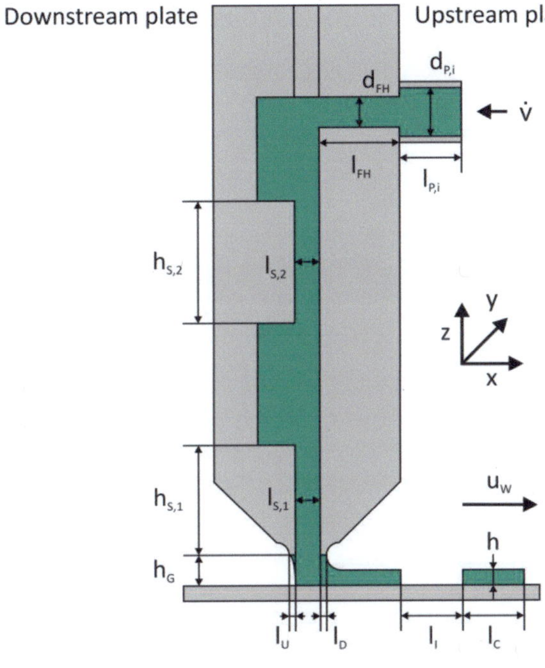

Figure 2.7: Schematic drawing of a slot die with all relevant dimensions in the height and length directions.

Table 2.2: Details of the three self-designed slot dies.

			Slot die 1	Slot die 2	Slot die 3
Slot die type			1-layer	2-layer	1-layer
Coating width	mm	$\mathrm{w}_{\mathrm{Coating}}$	60	60	12
FH length	mm	l_{FP}	20	10.75	20
FH diameter	mm	d_{FP}	4.5	4.5	4.5
Slot 1 height	µm	$\mathrm{h}_{\mathrm{S},1}$	500	500	500
Slot 1 length	mm	$\mathrm{l}_{\mathrm{S},1}$	25	25	18
Slot 2 height	µm	$\mathrm{h}_{\mathrm{S},2}$	500	-	500
Slot 2 length	mm	$\mathrm{l}_{\mathrm{S},2}$	25	-	20
Lip length D	mm	l_{D}	1	1	0.4
Lip length U	mm	l_U	1	1	1
Lip length M	mm	l_M	-	0.5	-

Slot die 1 was designed to investigate single-layer process windows. Its 60 mm wide coating width was chosen to be wide enough for an explicit visual observation of coating defects. Furthermore, this width is capable of a lab-scale electrode manufacture, although such an undertaking is beyond the scope of this work. In addition to the first cavity, slot die 1 has a second cavity, so as to be compatible with a wide range of viscosities. Their dimensioning was oriented to common available standards (Kistler and Schweizer, 1997; Russner, 2012). The lips machined into the end of the slot die plates have an overall edge-angle of 90° and are 1 mm long.

To investigate bilayer coatings, slot die 2 was designed with two feed slots. With only one cavity per feed slot, its dimensions are similar to the lower part of slot die 1, whereas the feed-holes are machined from the top into the cavities. The lips and slots are divided by an additional mid-lip with a length of 0.5 mm. The coating width is the same as for slot die 1.

To investigate coating edge effects and intermittent coatings, one requires knowledge of the total amount of cast liquid. This then allows one to easily identify a neck-in in coating width or spot liquid leakage during an interruption. To meet this requirement and the available sensor detection width (16 mm), 12 mm as the coating width for slot die 3 was chosen. Like slot die 1, slot die 3 also has two distribution cavities, but a smaller downstream lip length of only 0.4 mm. If desired, the second cavity can house a pressure transducer.

2.3.3 Continuous coatings

The basic experimental set-up was used to investigate process windows and coating edge effects. Before each experimental run, the desired slot die was mounted and the parameters (speed, gap, volume flow, etc.) were adjusted (Junges, 2011; Schneider, 2012; Bürkin, 2012; Wagner, 2012; Rolinger, 2013). To prevent misinterpretation of start-up phenomena, the coating was not analyzed until it reached a steady state film.

2.3.3.1 Process windows – Experimental procedure

During the investigations on process windows, slurry 1 and the Newtonian model system were coated with slot dies 1 and 2. For different volume flows, in the case of bilayer coating with slot die 2, a second identical syringe pump was used. The range of applied process parameters was selected to be in accordance with industrial practice. Thus, the coating speed ranged from 5 to 75 m/min, at a gap height of $h_G = 128\ \mu m$.

To determine the process limits, the operator visually characterized the film. If there were no stable coating conditions at a given set of process parameters, coating defects occurred (see Figure 1.6). Starting from stable coating conditions, the operator decreased the feed rate at a constant speed until the first defects occurred. This coating limit could then be correlated to the sum of adjusted gap and roller inaccuracy.

2.3.3.2 Coating edge effects – Experimental procedure

To investigate edge effects, slurry 2 was coated using the smaller slot die 3. Here, the steady state could be identified exactly by the height sensor system. It was reached when the actual film cross-sectional area matched a preset target value. Since, the theory implies an effect of the gap ratio, the gap and film heights were varied. Three different gap heights were used: 128, 101 and 76 µm. Then, for each gap height, the film height was reduced in 10 µm increments from an initial value close to the gap height until coating defects appeared. In further test series, the coating speed was varied to verify possible independence from the edge height. Here, the wet film thickness was held constant ($\mathbf{h} = 90$ µm for $\mathbf{h_G} = 128$ µm and 101 µm; $\mathbf{h} = 70$ µm for $\mathbf{h_G} = 76$ µm), while the coating speed was increased from 5 to 20 m/min.

Due to roller runout, every experimental trial was averaged for at least one turn of the roller. The recordings were then analyzed for bulk film height, edge heights, coated width and cross-sectional area. A schematic cross-sectional height profile is plotted in Figure 2.8.

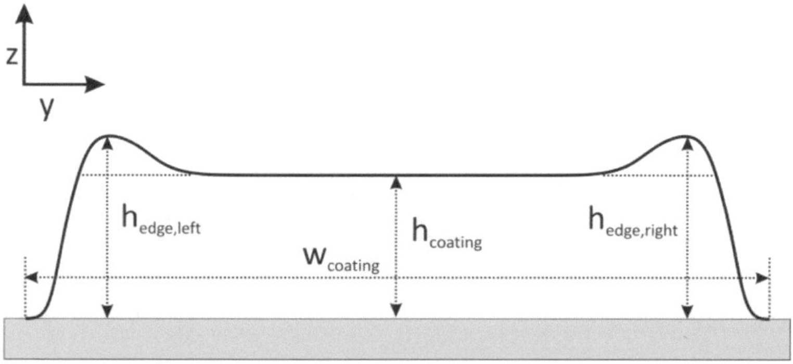

Figure 2.8: Schematic cross-sectional profile of a coated film with the relevant height and width dimensions.

2.3.4 Intermittent coatings

To study intermittent slot die coating and the related technologies, the basic experimental set-up had to be enhanced (Vogt, 2012; Kitz, 2013; Diehm, 2014). Care was taken to maintain transferability to most of the currently used industrial methods. Thus, the set-up included only the most essential equipment. The main difference to the basic arrangement is a bypass-valve located in front of the slot die. This valve directs the liquid flow either into the die or into a bypass waste pipe. The method explicitly excludes any process to prevent slurry leakage during the interruption state.

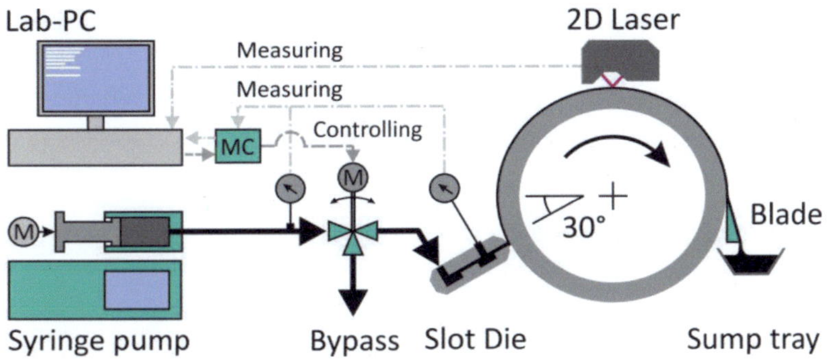

Figure 2.9: Schematic flow chart of the experimental set-up for intermittent coatings with control and measuring devices.

As with the basic set-up, an identical syringe pump with stiff two-piece syringes is used to supply liquid. It provides a constant, back pressure-independent volume flow, which was needed for the investigations in this work. A pumping system with a performance curve (e.g., progressing cavity pumps) might amplify back pressure variations caused by the switching bypass valve and complicate their measurement.

The pneumatic bypass valve (Swagelok, SS-43GXLS6MM-A15S3) is the key part of this set-up and there are many conceivable layouts. Again, the complexity was kept to a minimum by integrating a basic 3-way pneumatically-driven ball valve. The pressurized air rotating the bypass-valve was directed by magnetic valves (Festo, VOVG-S12-M32C-AH-M5-1H2) commanded by the microcontroller (MC). Switching to the waste-pipe direction was spring supported and therefore, faster than switching to the slot die direction. During each coating interruption, the liquid was directed through a waste pipe into a collection tank. During each coating phase, the bypass valve directed the liquid into the slot die. Just as for the coating edge investigations, slot die 3 was used here as well.

To analyze fluid pressure during the coating, two pressure transducers (Omega PR25Y-V-10-M12) were installed. The first, located between the syringe pump and the valve, detected any liquid pressure variation in the upstream system (p_{System}). The second was integrated in the lower cavity of the slot die (p_{Die}). To permit an on-line, time-resolved evaluation, the data was collected by a microcontroller (PJRC, Teensy 3.1) and sent to the operating PC.

The microcontroller was also programmed to control the bypass-valve via pressurized air valves. After receiving corresponding orders from an operating PC, the microcontroller switched into an endless intermitting loop. In dependence on the adjusted coating speed and the coating structure lengths, the controller then repeatently opened and closed the valve. Compared to issuing direct commands via the PC, this method was highly reproducible, since the MC was not interrupted by other tasks (Diehm, 2014). To exchange commands and data with the microcontroller, a LabView® (National Instruments, LabView® 2011) based program was developed. The program not only receives the measured liquid pressure, but also the state of the switching valve. This made it possible to measure the time elapsed between a valve-switch command and the liquid pressure reaction.

In addition, the film height data measured by the triangulation sensor was recorded separately on the operating PC. As for the coating edge investigations, the sensor records the whole film width, as shown in Figure 2.8.

To determine the web-directional height, only the central 6 mm of each profile were averaged. This avoided falsification of the height information by the coating edges. Due to the velocity dependent time delay between film application and height measurement at the roller top, a precise correlation between liquid pressure and resulting height was not possible. Nevertheless, the coated patterns were highly reproducible, which enabled for a congruent assignment of random patterns. Due to the correlation of film height, volume flow through the die and pressure drop, each change in trend of the observed film height should correspond to a change in die pressure.

A better process understanding might go in hand with knowledge of the relationship between system pressure and wet film thickness. Thus, various lengths of bypassing waste pipes were installed, in order to manipulate liquid pressure during the interruption phases. The recorded film dimensions, as plotted in Figure 2.10, were then compared to the measured pressure distributions. Furthermore, the limits in speed of the applied experimental set-up were analyzed by means of the above-mentioned reaction time.

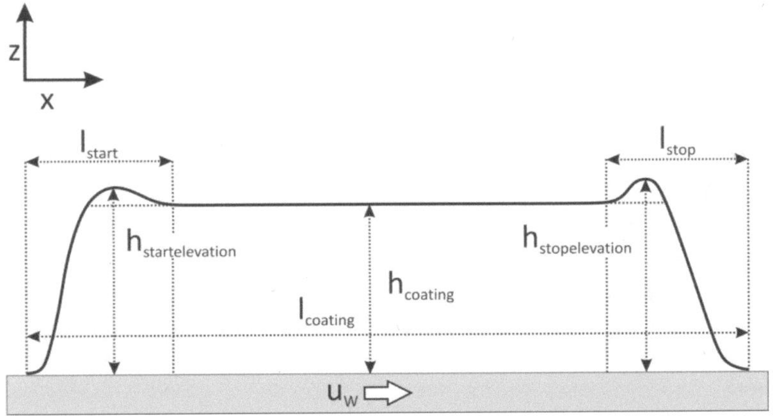

Figure 2.10: Schematic web-directional profile of a coated intermitted film with the relevant height and length dimensions.

The experimental procedure for the intermittent coating trials started by coating a continuous film. When a steady state was reached, as verified by the measured cross-sectional area, the intermitting process was started. The liquid pressure distribution and film height were then recorded for at least one complete rotation of the roller.

3 Process windows

Homogeneous coatings can only be obtained within certain coating limits. The limiting forces are defined by the bead pressure balance and the low-flow limit. As described in 1.2.1, they are well known and computable, even for non-newtonian liquids.

To obtain information about the stability of the coating bead, numerical simulations can be performed by solving the two-dimensional Navier-Stokes equations (Sartor; Gates, 1999; Carvalho and Kheshgi, 2000; Romero, 2004; Lee et al., 2011). However, the more precise these simulations are, the more time it takes to generate them and prepare the results. The complexity of such flow simulations is further increased for viscoelastic polymer solutions that include large particles, such as lithium-ion battery coatings.

Along with these simulations, the available analytical methods mentioned n section 1.2.1.1 are used to predict the slot die process window. A main advantage here is that the computation of these one-dimensional analytical expressions proceeds promptly.

Nevertheless, the impact on the analytical models of combining relatively high viscosities with particle and polymer loadings remains still unknown. In chapter 3.1, these models will be applied to predict the coating limit for a single layer LIB coating. Further, the calculations will be discussed and compared to experimental results.

Although the application of bilayer coatings offers certain economic advantages, it also entails certain phenomena that must be managed (see chapter 1.2.2). To examine the characteristics of the altered gap region, the single-layer models will be extended and discussed in chapter 3.2.

3.1 Single layer coatings

Within this chapter the available models for single layer coating windows will be applied and discussed. To validate the available coating window models for battery slurries, both, model predictions and experimental coatings will be accompanied by those for a Newtonian liquid to reveal rheological singularities.

3.1.1 Film break-up models

As already briefly shown illustrated in chapter 1.2.1, a closed film coating is only possible within stable, stationary menisci. The position and stability of the upstream meniscus depends on a coating bead pressure balance. This limit can be extended by shifting the pressure drop with a vacuum chamber on the upstream side. In contrast, the low-flow limit determines the stability of the downstream meniscus and marks the final minimum film thickness.

It could be experimentally shown, that the most serious process defects in LIB coatings at ambient pressure conditions are caused by air entrainment (Schmitt et al., 2013a) as to see in Figure 1.6. A model to predict the minimum film thickness or maximum speed for LIB coatings will focus on air entrainment rather than low flow limits. The model will be applied for Newtonian liquids, to show the impact of the shear-dependent slurry behavior.

The basis of this model is provided by the pressure distribution calculations in the coating gap. Regarding this coating gap, the feed slot divides a coating gap into downstream and upstream sections, as presented in Figure 3.1.

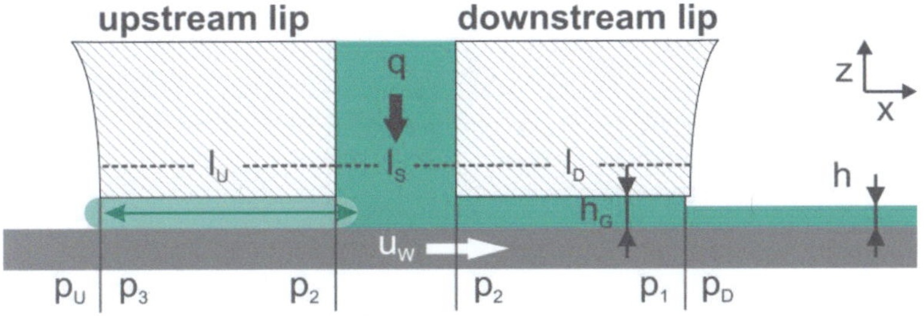

Figure 3.1: Schematic drawing of a slot die gap region showing all parameters relevant for bead stability.

The overall coating bead pressure gradient Δp_{DU} (**D**ownstream/**U**pstream) is determined by the individual sections, including the capillary pressure at the interfaces (Higgins and Scriven, 1980; Durst and Wagner, 1997; Durst et al., 2007):

$$\Delta p_{DU} = p_D - p_U = \Delta p_{D1} + \Delta p_{12} + \Delta p_{23} + \Delta p_{3U} \tag{3.1}$$

Within the downstream section, there is a superposition of the web-driven Couette flow and the pressure-induced Poiseuille flow. For Newtonian liquids, the pressure difference in this section is given by (Durst et al., 2007):

$$\Delta p_{12} = \left(\frac{u_W h_G}{2} - q\right)\frac{12\,\eta\,l_D}{h_G^3} \tag{3.2}$$

with u_W as the coating speed, h_G as the gap height, q as the specific volume flow, η as the viscosity and l_D as the length of the downstream lip. Since it is reasonable to relate film stability to a certain wet film thickness h, Durst (2007) substituted the volume flow in (3.2) by equation (1.1):

$$\Delta p_{12} = \left(1 - \frac{2h}{h_G}\right)\left(\frac{6\,\eta\,u_W\,l_D}{h_G^2}\right) \tag{3.3}$$

In cases where the upstream section is stationary wetted, there must be a circulation, which leads to a zero net flow. Thus, q in (3.2) becomes zero and the pressure difference inside the upstream section for Newtonian flows is given by (Durst et al., 2007):

$$\Delta p_{23} = \left(\frac{6\,\eta\,u_W\,l_U}{h_G^2} \right) \tag{3.4}$$

were l_U is the wetted length of the upstream lip. For the liquid-gas interface at the downstream edge, Ruschak (1976) applied an expression for the capillary pressure within a film-forming surface to:

$$\Delta p_{D1} = 1.34\,Ca^{\frac{2}{3}} \frac{\sigma}{h} \tag{3.5}$$

With the capillary number $\boldsymbol{Ca}$ and the liquid surface tension $\boldsymbol{\sigma}$. This expression was developed from a numerical approximation for film-foming menisci in a Newtonian drag coating[2] from Landau and Levich (1942). The capillary number is given by:

$$Ca = \frac{\eta \cdot u_W}{\sigma} \tag{3.6}$$

At the upstream meniscus, a simple capillary pressure gradient is often assumed, in dependency on the gap height. Since the meniscus could be either convex or concave, this term is either positive or negative (Durst et al., 2007):

$$\Delta p_{3U} = \pm \frac{2\sigma}{h_G} \tag{3.7}$$

There are even more detailed models available for both liquid-gas interfaces (Higgins and Scriven, 1980; Weinstein and Ruschak, 2004). However, since their impact is small (more details shown in Appendix 8.8) and

[2] *Continuous dip coating by a submerged moving web.*

they require more process information, the literature and this work tend to remain with the simplified expressions (Yang et al., 2004b; Youn et al., 2006; Durst et al., 2007; Lee et al., 2011; Koh et al., 2012; Jang and Song, 2013).

Due to the four possible locations of the upstream meniscus, as described in chapter 1.2.1.1, the combination of equations (3.3)-(3.7) in (3.1) results in two pairs of pressure balance-limits (Durst et al., 2007):

Case 1 $$\Delta p_{DU} \leq 1.34\, Ca^{\frac{2}{3}} \frac{\sigma}{h} + \left(1 - \frac{2h}{h_G}\right)\left(\frac{6\,\eta\, u_W\, l_D}{h_G^2}\right) + \left(\frac{6\,\eta\, u_W\, l_U}{h_G^2}\right) + \frac{2\sigma}{h_G} \tag{3.8}$$

Case 2 $$\Delta p_{DU} \geq 1.34\, Ca^{\frac{2}{3}} \frac{\sigma}{h} + \left(1 - \frac{2h}{h_G}\right)\left(\frac{6\,\eta\, u_W\, l_D}{h_G^2}\right) + \left(\frac{6\,\eta\, u_W\, l_U}{h_G^2}\right) - \frac{2\sigma}{h_G} \tag{3.9}$$

Case 3 $$\Delta p_{DU} \leq 1.34\, Ca^{\frac{2}{3}} \frac{\sigma}{h} + \left(1 - \frac{2h}{h_G}\right)\left(\frac{6\,\eta\, u_W\, l_D}{h_G^2}\right) + \frac{2\sigma}{h_G} \tag{3.10}$$

Case 4 $$\Delta p_{DU} \geq 1.34\, Ca^{\frac{2}{3}} \frac{\sigma}{h} + \left(1 - \frac{2h}{h_G}\right)\left(\frac{6\,\eta\, u_W\, l_D}{h_G^2}\right) - \frac{2\sigma}{h_G} \tag{3.11}$$

Here case 1 and 2 represent a convex and a concave mensicus located at the outer edge of the upstream lip. If the pressure increases in comparison to case 1, die swell occurs. In contrast, case 3 and 4 represent a convex and a concave upstream-meniscus located at the inner edge of the upstream lip. If the pressure decreases beneath case 4, air entrainment occurs.

Since LIB coatings show a shear-thinning behavior, it is reasonable to include a non-Newtonian viscosity in the process window model. In the relevant shear-range, the viscosity of LIB coatings can be described by a power-law approach:

$$\eta = \kappa\, \dot{\gamma}^{\,\varepsilon - 1} \tag{3.12}$$

with the consistence factor κ, the power-law exponend ε and the shear rate $\dot{\gamma}$. In the coating gap the shear rate is approximately described by:

$$\dot{\gamma} = \frac{u_W}{h_G} \tag{3.13}$$

Tsuda (2010) and Lee (2011) solved the resulting complex equation for motion of fluids using the power-law approach. For the single-layer coating bead, Lee describes two expressions for the downstream pressure gradient (Lee et al., 2011):

For $h < h_G/2$ $$\Delta p_{12} = l_D \cdot \kappa \cdot \left(\frac{(\varepsilon+1)(2\varepsilon+1)}{\varepsilon}\right)^{\varepsilon} \cdot \frac{(h_G - 2 \cdot h)^{\varepsilon}}{h_G^{(2\varepsilon+1)}} \cdot u_W^{\ \varepsilon} \tag{3.14}$$

For $h > h_G/2$ $$\Delta p_{12} = -\, l_D \cdot \kappa \cdot \left(\frac{(\varepsilon+1)(2\varepsilon+1)}{\varepsilon}\right)^{\varepsilon} \cdot \frac{(2 \cdot h - h_G)^{\varepsilon}}{h_G^{(2\varepsilon+1)}} \cdot u_W^{\ \varepsilon} \tag{3.15}$$

As in equation (3.3), the wet film thickness in the downstream pressure drop represents the net flow in the area under consideration. For the upstream section, the net flow of zero again results in a differing expression:

$$\Delta p_{23} = l_U \cdot \kappa \cdot \left(\frac{(\varepsilon + 1)(2\varepsilon + 1)}{\varepsilon}\right)^{\varepsilon} h_G^{-(\varepsilon+1)} \cdot u_W^{\ \varepsilon} \tag{3.16}$$

Describing the capillary pressure for the downstream meniscus, the capillary number is now a function of the shear rate (Kamişli and Ryan, 1999). The combination of equations (3.6), (3.12) and (3.13) solves the capillary number for liquids following the power-law approach:

$$Ca = \kappa \frac{u_W^{\varepsilon}}{\sigma \cdot h_G^{\varepsilon - 1}} \tag{3.17}$$

Although Ruschak's expression in (3.5) was derived for Newtonian liquids, only minor deviations were expected in literature. Moreover, for thicker films, the size of the downstream meniscus pressure drop has a relatively low impact.

Combining the Newtonian equations (3.8)-(3.11) with (3.15)-(3.17) the four stability cases of a power-law coating bead for the example of $h > h_G/2$ are:

Case 1
$$\Delta p_{DU} \leq 1.34\left(\kappa\frac{u_W^{\varepsilon}}{\sigma\cdot h_G^{\varepsilon-1}}\right)^{\frac{2}{3}}\frac{\sigma}{h} + l_D\cdot\kappa\cdot\left(\frac{(\varepsilon+1)(2\varepsilon+1)}{\varepsilon}\right)^{\varepsilon}\cdot\frac{(2\cdot h-h_G)^{\varepsilon}}{h_G{}^{(2\varepsilon+1)}}\cdot u_W{}^{\varepsilon}$$
$$+ l_U\cdot\kappa\cdot\left(\frac{(\varepsilon+1)(2\varepsilon+1)}{\varepsilon}\right)^{\varepsilon} h_G^{-(\varepsilon+1)}\cdot u_W{}^{\varepsilon} + \frac{2\sigma}{h_G} \tag{3.18}$$

Case 2
$$\Delta p_{DU} \geq 1.34\left(\kappa\frac{u_W^{\varepsilon}}{\sigma\cdot h_G^{\varepsilon-1}}\right)^{\frac{2}{3}}\frac{\sigma}{h} + l_D\cdot\kappa\cdot\left(\frac{(\varepsilon+1)(2\varepsilon+1)}{\varepsilon}\right)^{\varepsilon}\cdot\frac{(2\cdot h-h_G)^{\varepsilon}}{h_G{}^{(2\varepsilon+1)}}\cdot u_W{}^{\varepsilon}$$
$$+ l_U\cdot\kappa\cdot\left(\frac{(\varepsilon+1)(2\varepsilon+1)}{\varepsilon}\right)^{\varepsilon} h_G^{-(\varepsilon+1)}\cdot u_W{}^{\varepsilon} - \frac{2\sigma}{h_G} \tag{3.19}$$

Case 3
$$\Delta p_{DU} \leq 1.34\left(\kappa\frac{u_W^{\varepsilon}}{\sigma\cdot h_G^{\varepsilon-1}}\right)^{\frac{2}{3}}\frac{\sigma}{h} + l_D\cdot\kappa\cdot\left(\frac{(\varepsilon+1)(2\varepsilon+1)}{\varepsilon}\right)^{\varepsilon}\cdot\frac{(2\cdot h-h_G)^{\varepsilon}}{h_G{}^{(2\varepsilon+1)}}\cdot u_W{}^{\varepsilon} + \frac{2\sigma}{h_G} \tag{3.20}$$

Case 4
$$\Delta p_{DU} \geq 1.34\left(\kappa\frac{u_W^{\varepsilon}}{\sigma\cdot h_G^{\varepsilon-1}}\right)^{\frac{2}{3}}\frac{\sigma}{h} + l_D\cdot\kappa\cdot\left(\frac{(\varepsilon+1)(2\varepsilon+1)}{\varepsilon}\right)^{\varepsilon}\cdot\frac{(2\cdot h-h_G)^{\varepsilon}}{h_G{}^{(2\varepsilon+1)}}\cdot u_W{}^{\varepsilon} - \frac{2\sigma}{h_G} \tag{3.21}$$

As for the Newtonian liquids described above, continuous films of non-Newtonian coatings are possible between case 1 and case 4. If the coating bead pressure drops below case 4 (equation (3.21)), then air entrainment limits a stable film application.

For a further investigation of coating defects it is reasonable to compare the above air entrainment model with the low-flow limit, represented by the visco-capillary model (VCM).

It is based on the primary purpose of equation (3.5), which is to define the dependence of liquid parameters and pressure drop to the stability of a film-forming meniscus (Landau and Levich B., 1942; Ruschak, 1976; Carvalho and Kheshgi, 2000). For a relation of limiting coating parameters and meniscus stability, this equation can be related to the capillary pressure:

$$p_D - p_1 = \frac{\sigma}{r} \tag{3.22}$$

Assuming a cylindrical meniscus shape, the radius r can be substituted for by the gap and film heights:

$$r = \frac{h_G - h}{2} \tag{3.23}$$

Combining , (1.2), (3.10), (3.22) and (3.23) leads to the visco-capillary model, where a maximum dimensionless gap ratio is related to a critical capillary number:

$$Ca_{Crit} = N\left(\frac{2}{G_{Max}-1}\right)^{\frac{3}{2}};N = 0.65 \tag{3.24}$$

Here N is a numerical based factor. To match the model to his experimental results Wengeler (2014) adapted this factor to $N_{Wengeler} = 0.128$. The VCM shows an increase in minimum film thickness towards higher capillary numbers, as shown in Figure 1.9.

Combining the above air entrainment models and the visco-capillary model, Figure 3.2 and Figure 3.3 show the process windows for the Newtonian model system and slurry 1 for various coating speeds. The calculations are based on the liquid parameters in Table 3.1, Slot die 1, $h_G = 148\ \mu m$ and $h = 80\ \mu m$.

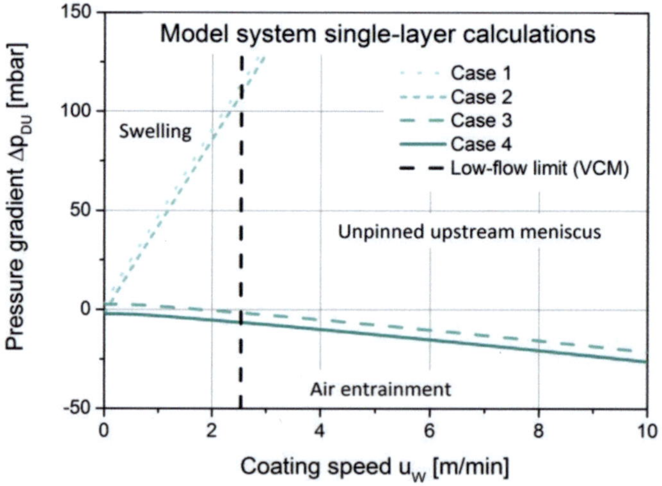

Figure 3.2: The calculated process window for the Newtonian model system at a wet film thickness of $h = 80\ \mu m$ *and a gap of* $h_G = 148\ \mu m$.

The coating window for the Newtonian model system in Figure 3.2 follows a form that is typical for high viscosities and relatively low surface tension. As implied by equations (3.8)-(3.11), the viscosity describes the spreading of the side warded approximately triangular area, which would be narrower for lower viscosities. In contrast, the gap ratio of $G = 1.85$ affects the whole orientation of the triangle. It would be oriented towards positive pressure gradients for smaller film heights or higher gaps.

As mentioned in 1.2.1.1, the areas where the upstream meniscus is stationary and stable are between Cases 1 and 2, or Cases 3 and 4. The group of linear pairs describes an opening triangular area by departing from each other towards increasing coating speeds. In this area, between Case 2 and 3, the upstream meniscus is not stably pinned.

In concrete terms, for $\Delta p_{DU} = 0$ (as for all experiments performed in this work), the upstream meniscus is only pinned at the inner edge of the upstream lip until $u_w \approx 1.5\, m/min$. Beyond this speed, the coating bead is fragile in the face of external disturbances. Nevertheless, the only theoretical border regarding higher speeds is defined by the low-flow limit at $u_W = 2.5\, m/min$. In previous works, it was shown, that it is possible to coat LIB electrodes films inside this "unpinned" area even at higher speeds (Schmitt et al., 2013a; Schmitt et al., 2013c). Although, the oscillating upstream meniscus provokes uneven film heights it cannot be seized as a definite break-up mechanism. Remarkably the most interesting graph—Case 4—representing air entrainment, is orientated in negative direction. Thus, air entrainment will not appear for these conditions, even at higher coating speeds and the coating has to be limited by another mechanism.

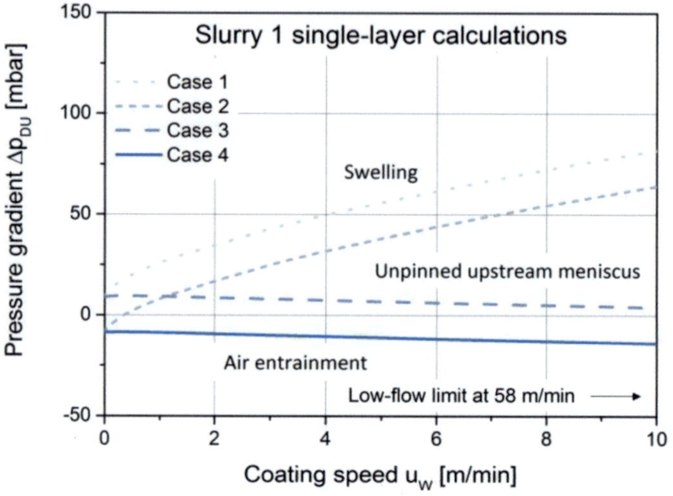

Figure 3.3: The calculated process window for Slurry 1 at a wet film thickness of $h = 80\ \mu m$ and a gap of $h_G = 148\ \mu m$.

The power-law coating bead pressure balances (see Figure 3.3) describe similar, but non-linear, coating limits. For lower coating speeds, there is a slight curvature. This can be attributed to the bigger contribution of Δp_{D1} in this region. Because an exponent smaller than 1 is selected for the speed in the power-law calculations, the impact of this pressure gradient decreases for increasing speeds. The more conspicuous difference to Figure 3.2 is given by the increased coatable range in the ordinate direction. This is due to the larger liquid surface tension, which can be seen at $u_W = 0\ m/min$. Here, equation (3.1) is reduced to the capillary pressure of $\pm\Delta p_{3U}$.

Again, case 4 will not be reached by an increase in speed at ambient conditions. The coating is theoretically limited only by a low-flow. With a capillary number, given in (3.6), the VCM predicts the low-flow limit to be $u_W = 28.8\ m/min$. This limiting speed is far larger than for the Newtonian case, due to the changed liquid properties. While the surface tension increases more than threefold, the viscosity decreases dynamically within several magnitudes towards higher speeds. Hence, u_W has less impact on the capillary number Ca of shear-thinning slurry 1 than it does on the capillary number of the Newtonian model system. As shown in Figure 1.9, this results in higher coating speeds at the same G and Ca.

For LIB coatings, air entrainment has been reported as the prevailing limiting mechanism for decreasing wet film thicknesses at given speeds (Kaiser et al., 2014). To characterize its onset, it makes sense to investigate the behavior of case 4 for decreasing film heights. The results of such an investigation are plotted for various coating speeds and both liquids in *Figure 3.4* and *Figure 3.5*.

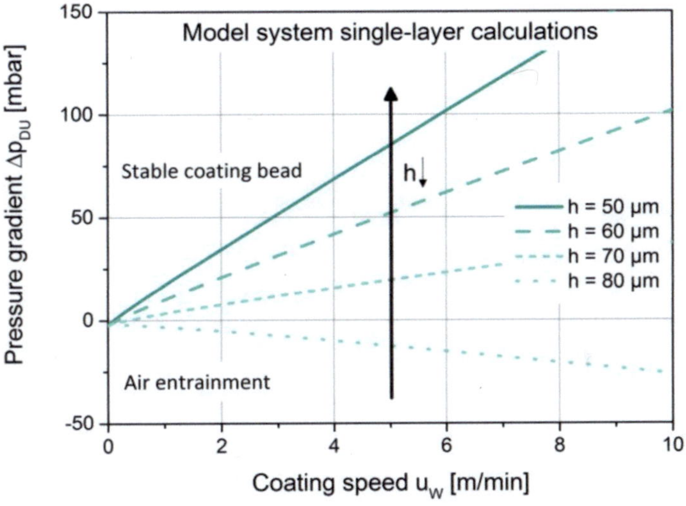

Figure 3.4: The calculated case 4, notably the onset of air entrainment, for the Newtonian model system at a gap of $h_G = 148\,\mu m$ and several film heights.

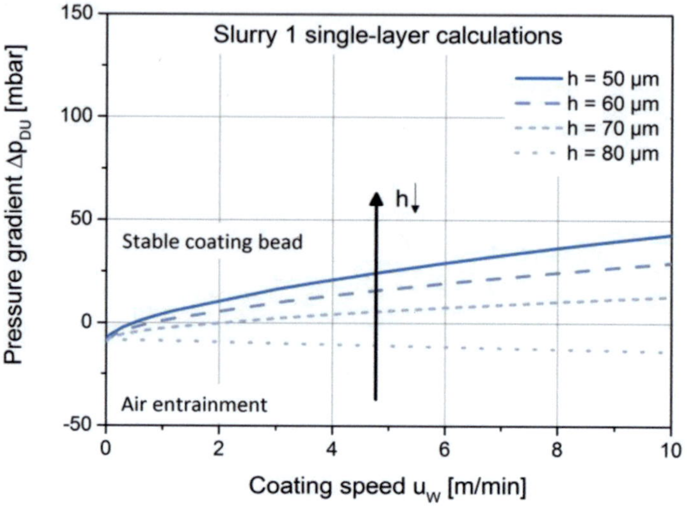

Figure 3.5: The calculated case 4, notably the onset of air entrainment, for slurry 1 at a gap of $h_G = 148\ \mu m$ and several film heights.

As mentioned above, the orientation of the whole coating window turns towards positive pressure gradients with decreasing film heights. With special regard to case 4, this results in a crossing of the zero pressure line. Here, the upstream meniscus stretches beyond its stable dimensions. Thus, it breaks up and air entrainment occurs. Comparing the calculations in *Figure 3.4* and *Figure 3.5*, the power-law calculations are again more curved and, due to the higher surface tension, start at a lower pressure difference.

Case 4 crosses the zero line for a film thickness of $h \approx h_{gap}/2$. Here, Δp_{12} changes its sign and both capillary pressures Δp_{D1} and Δp_{3U} are approximately equal. In the above examples, for $h = 70\ \mu m$ and $\Delta p_{DU} = 0\ mbar$, air entrainment would occur at $u_W = 0.3\ m/min$ for the model system and at $u_W = 2\ m/min$ for slurry 1. For every combination of pressure difference and film thickness, there is an associated maximum coating speed.

Thus, being interested in identifying the minimum film thickness as a function of speed, case 4 must be calculated implicitly. To compare the calculations with experimental results made without a vacuum chamber, the pressure gradient was set to zero. A solver then computed the minimum film thickness as a function of coating speed. The computed minimum film heights for the Newtonian and power-law calculations are plotted in Figure 3.6 and Figure 3.7.

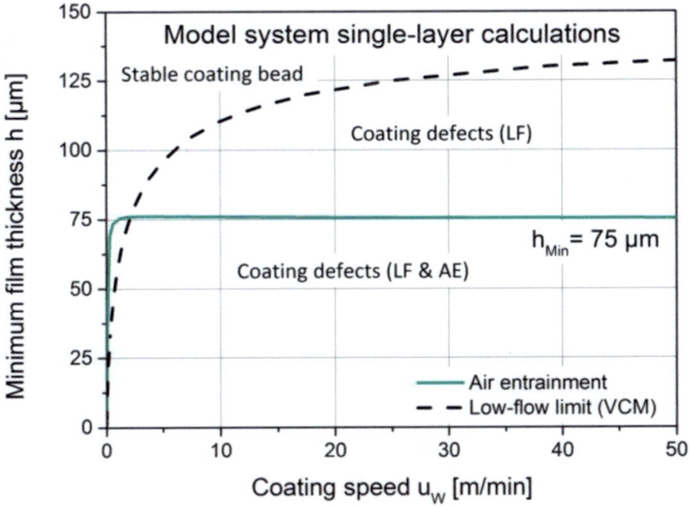

Figure 3.6: Calculated minimum film thickness in dependence on the onset of air entrainment and the low-flow limit for Newtonian model system at $h_G = 148\ \mu m$.

The Newtonian calculations (see Figure 3.6) show a horizontal distribution of the minimum film thickness beyond speeds of $u_W \approx 2\ m/min$. Thus, a film thicker than the computed $h_{Min} = 75\ \mu m$ should not be limited by air entrainment. As mentioned above, the value of h_{Min} is based on the change in sign of Δp_{12} for $h = h_G/2$. The slight deviation is due to the contribution of the menisci, which has to be eliminated to fulfill $\Delta p_{DU} = 0$. At speeds lower than $u_W = 2\ m/min$, the contribution of the motional pressure gradient decreases relative to the capillary upstream pressure Δp_{3U}. Thus, the limit develops a curvature towards the zero point, and thinner films are possible in this area.

In comparison, the low-flow limit, represented by the VCM in (3.24), describes thicker minimum film thicknesses. The graph already crosses the theoretical onset of air entrainment at $u_W = 2\ m/min$. Thus, beyond this speed, the low-flow limit should be the prevailing coating defect for the Newtonian model system.

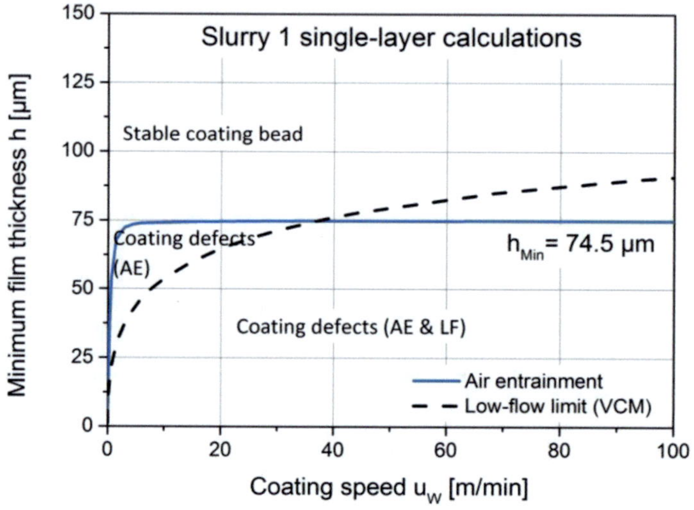

Figure 3.7: Calculated minimum film thickness in dependence on the onset of air entrainment and the low-flow limit for slurry 1 at $h_G = 148\ \mu m$.

For the power-law calculations, the minimum film thickness reaches its horizontal gradient below $u_W = 0.5\ m/min$. This is due to the higher value of the speed-dependent pressure gradients at lower speeds relative to the capillary upstream pressure. For the same reason, the minimum film thickness decreases to $h_{Min} = 74.5\ \mu m$.

As already presented in Figure 3.3, the low-flow limit occurs for slurry 1 at higher speeds, compared to the model system. It does not cross the onset of air entrainment until $u_W = 35\ m/min$. Beyond this point, the related film thickness increases more slowly than observed in Figure 3.6. According to these calculations, coated films of slurry 1 that are thicker than $h \approx 80\ \mu m$ should not be limited until $u_W = 50\ m/min$. This fits the needs of thick LIB electrode coatings with thicknesses of $h \approx 100\ \mu m$ or higher (Neumann, 2011).

3.1.2 Experimental results and discussion

To validate the model of LIB coating process windows, model predictions had to be compared to experimental results. Here, the basic experimental set-up, as described in chapter 2.3, was used. The coating limit was characterized by reducing the applied film thickness at a fixed speed. The minimum film thickness was reached when the first coating defects visually appeared in the coating, as shown in *Figure 3.8*.

Figure 3.8: Stable LIB-electrode coating (left) and onset of film defects for a reduced film thickness (right) at the used experimental set-up.

The liquid data of the coated slurry 1 and the Newtonian model system are listed in Table 3.1.

Table 3.1: Power-law parameters, viscosity and surface tensions of the slurry 1 and the Newtonian model system.

	Slurry 1	Model system	
κ	3.2	-	[Pa s$^{\varepsilon}$]
ε	0.67	-	[-]
η	-	1	[Pa s]
σ	0.066	0.018	[N/m]

In *Figure 3.9* and *Figure 3.10*, the experimental results of the slurry and the model system are plotted together with model predictions.

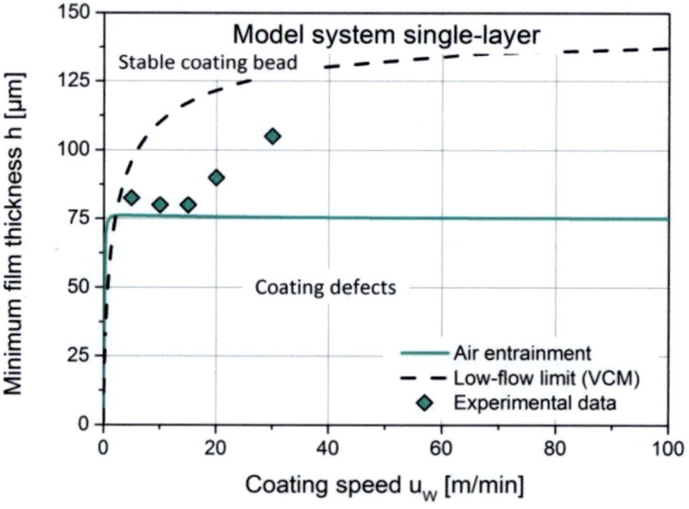

Figure 3.9: Predicted and observed minimum film thickness of the Newtonian model system at $h_G = 148\ \mu m$ (Schmitt et al., 2015b).

For the Newtonian model system, the observed film break-ups occured as air entrainment defects in cross-web direction. They correspond to the prediction of the air entrainment model until $u_W = 15\ m/min$. For $h_{Min} \approx 80\ \mu m$, the results are only a little higher than the predicted theoretical limit. This might be attributable to external disturbances, such as fluctuations in gap height or volume flow. For higher coating speeds beyond $15\ m/min$, there is a vivid change in trend of the observable minimum film thicknesses. They increase and narrow the predicted VCM break-up line. Experimental results for coating speeds beyond $30\ m/min$ were not possible with the presented set-up due to the high pressure drop of the Newtonian liquid towards the required feed rates.

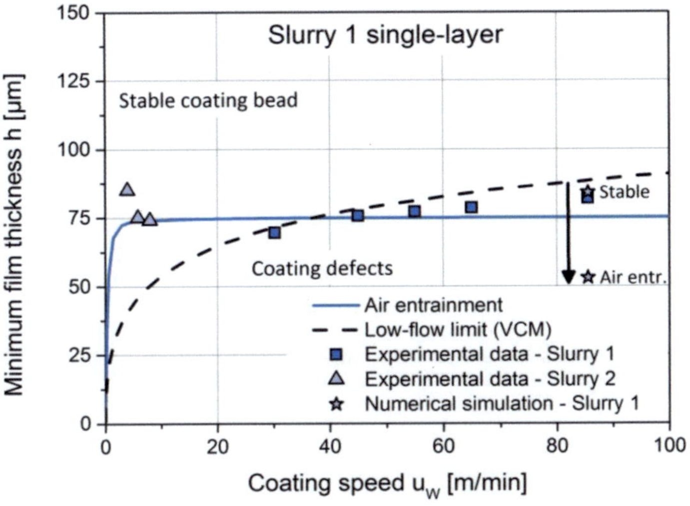

Figure 3.10: Predicted and observed minimum film thickness of slurry 1 (and 2) at $h_G = 148\ \mu m$ (Schneider, 2012; Wittek, 2012; Schmitt et al., 2013b; Schmitt et al., 2013a; Weiß, 2013).

Compared to the Newtonian calculations, the electrode coatings should be limited by air entrainment until $u_W \approx 40\ m/min$. Beyond this coating speed the low-flow limit should be the prevailing break-up mechanism.

Affirming these calculations, the experimentally observed break-up speeds of slurry 1 at relatively high speeds, line up in a row close to the visco-capillary model. Nevertheless, the occured coating defects were observed to be caused by upstream air entrainment. Additional numerical simulations (see Appendix 8.6) confirmed this observation (Wittek, 2012; Weiß, 2013). This might be attributable to low-flow disturbances by the downstream meniscus which do not yet provoke a film break-up but induce an upstream air entrainment.

However, the VCM was derived for low capillary number conditions. Even for the smallest coating speeds, they are beyond unity, with $Ca = 2.5$ for slurry 1 and $Ca = 4.6$ for the model system. Thus, and due to the fact that low-flow defects are not the present break-up phenomena, the VCM is not capable of predicting the process window of LIB electrodes. Other visco-capillary models, like those of Higgins (Higgins and Scriven, 1980) and Wengeler (Wengeler, 2014), would overestimate the results even more.

The observation of Carvalho and Kheshgi (2000) might deliver an explanation for the failed prediction by the VCM. As discussed in chapter 1.2.1.2 they observed an again falling low-flow limit towards higher capillary numbers of $Ca > 1$. Thus the high viscous coating of LIB-electrode slurries and the Newtonian model system would only be limited by the air entrainment.

The observed rising trend of minimum film thickness for slurry 1 therefore has to be caused by another mechanism. Mechanical disturbances, e.g. the staggering roller, can be one reason wich increasingly destabilize the coating bead towards higher coating speeds.

Unfortunately, low-speed coating trials were not done before the change in slurry receipe. Thus, there are no information about the break-up behavior of slurry 1 in the lower speed range. However, the viscosity data of slurry 2 (see Table 4.1) does not change the predicted onset of air entrainment dramatically ($\Delta h_{Min} = 1\ \mu m$). The available low-speed break-up results of slurry 2 are therefore comparable and are plotted in *Figure 3.10*. Similarly for slurry 1 at higher speeds, the low-speed film break-up of slurry 2 matches the air entrainment prediction. Microscopic image acquisition of the gap region made at these break-up conditions showed an overstreched upstream meniscus (as shown in appendix 8.7). This further comfirms the above discussed model for predicting air entrainment during the slot die coating of LIB electrodes. However, the calculations and experimental results showed that coatings of $h = 100\ \mu m$ are possible at $u_W = 100\ m/min$ without a film break-up. A dimensionless presentation of the above discussed results is given by *Figure 3.11*.

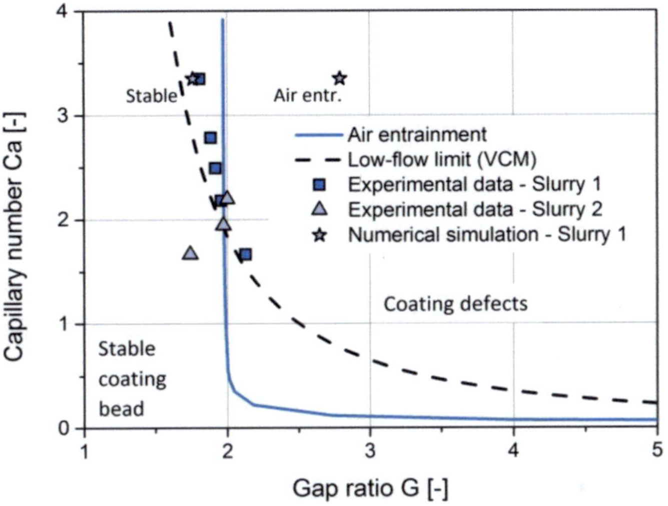

Figure 3.11: *Dimensionless plot of the predicted and observed minimum film thickness of slurry 1 (and 2) at $h_G = 148\ \mu m$ (Schneider, 2012; Wittek, 2012; Schmitt et al., 2013b; Schmitt et al., 2013a; Weiß, 2013).*

As presented in chapter 1.2.1.2 the process window in *Figure 3.10* can be plotted dimensionless. While the minimum wet film thickness is substituted by the gap ratio, the break-up web speed is replaced by the capillary number. The obvious differences to *Figure 3.10* are the matching break-up numbers of slurry 1 and 2. Due to their differing shear-thinning behavior, expressed in the consistence parameter κ and the exponent ε, the capillary number in equation (3.17) is less effected by the higher web speeds of the slurry 1 trials.

3.2 Bilayer coatings

For multilayer slot dies, it is reasonable that the altered gap regime also highly affects the processing limits. An adaptation of the models for single-layer process windows to the bilayer case might help to characterize its singuilarities. Therefore the available equations for Newtonian and shear thinning coating liquids will be extended, discussed and compared to experimental results in the following.

3.2.1 Model derivation

When deriving the four stability cases for bilayer coatings, all the singularities associated with bilayer coatings must be accounted for. A schematic sketch of a bilayer slot die gap with all relevant parameters is shown in Figure 3.12.

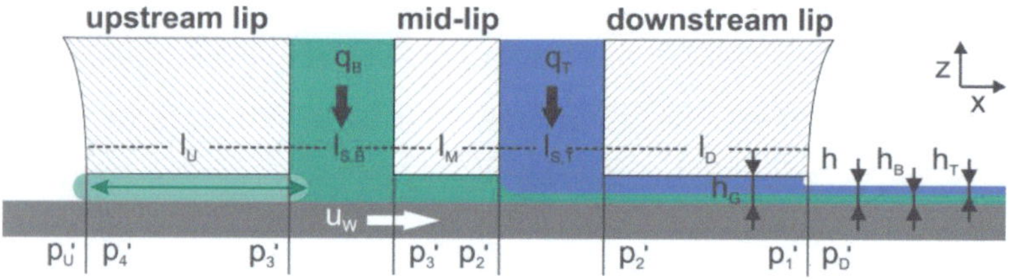

Figure 3.12: Schematic drawing of a bilayer slot die gap region with all parameters relevant for bead stability (Schmitt et al., 2015b).

Because the slot die process is pre-metered, the wet film thickness must equal the specific feed volume. For bilayer slot die coating, the resulting wet film thickness arises from the combination of top and bottom layer feed and thus their film heights:

$$h = \frac{q_B + q_T}{u_W} = h_B + h_T \tag{3.25}$$

In order to substitute the individual film heights, a top to bottom layer feed ratio $\boldsymbol{M}$ is introduced:

$$M = \frac{h_T}{h_B} \tag{3.26}$$

A new pressure gradient $\boldsymbol{\Delta p'_{23}}$ beneath the mid-lip has to be taken into account to describe the coating bead pressure $\boldsymbol{\Delta p'_{DU}}$ for bilayer applications. Here, the net flow is reduced to the specific bottom layer flow rate q_B. With a substitution of the flow rate by (3.25) and (3.26), the pressure difference underneath the mid-lip for Newtonian liquids becomes:

$$\Delta p'_{23} = \left(1 - \frac{2h}{h_G} \frac{1}{(1+M)}\right)\left(\frac{6\,\eta\,u_W\,l_M}{h_G^2}\right) \tag{3.27}$$

Compared to the single layer application in equation (3.1), the summarized pressure drop for bilayer slot die coatings must be extended to:

$$\Delta p'_{DU} = \Delta p'_{D1} + \Delta p'_{12} + \Delta p'_{23} + \Delta p'_{34} + \Delta p'_{4U} \tag{3.28}$$

Analogous to the single-layer equations (3.8)-(3.11), the two pairs of stability limits are:

Case 1
$$\Delta p'_{DU} = 1.34\,Ca^{\frac{2}{3}}\frac{\sigma}{h} + \left(1 - \frac{2h}{h_G}\right)\left(\frac{6\,\eta\,u_W\,l_D}{h_G^2}\right) + \left(1 - \frac{2h}{h_G}\frac{1}{(1+M)}\right)\left(\frac{6\,\eta\,u_W\,l_M}{h_G^2}\right) + \left(\frac{6\,\eta\,u_W\,l_U}{h_G^2}\right) + \frac{2\sigma}{h_G} \tag{3.29}$$

Case 2
$$\Delta p'_{DU} = 1.34\,Ca^{\frac{2}{3}}\frac{\sigma}{h} + \left(1 - \frac{2h}{h_G}\right)\left(\frac{6\,\eta\,u_W\,l_D}{h_G^2}\right) + \left(1 - \frac{2h}{h_G}\frac{1}{(1+M)}\right)\left(\frac{6\,\eta\,u_W\,l_M}{h_G^2}\right) + \left(\frac{6\,\eta\,u_W\,l_U}{h_G^2}\right) - \frac{2\sigma}{h_G} \tag{3.30}$$

Case 3 $$\Delta p'_{DU} = 1.34\, Ca^{\frac{2}{3}} \frac{\sigma}{h} + \left(1 - \frac{2h}{h_G}\right)\left(\frac{6\,\eta\, u_W\, l_D}{h_G^2}\right) + \left(1 - \frac{2h}{h_G}\frac{1}{(1+M)}\right)\left(\frac{6\,\eta\, u_W\, l_M}{h_G^2}\right) + \frac{2\sigma}{h_G} \tag{3.31}$$

Case 4 $$\Delta p'_{DU} = 1.34\, Ca^{\frac{2}{3}} \frac{\sigma}{h} + \left(1 - \frac{2h}{h_G}\right)\left(\frac{6\,\eta\, u_W\, l_D}{h_G^2}\right) + \left(1 - \frac{2h}{h_G}\frac{1}{(1+M)}\right)\left(\frac{6\,\eta\, u_W\, l_M}{h_G^2}\right) - \frac{2\sigma}{h_G} \tag{3.32}$$

In analogy with the Newtonian calculation for coatings described by power-law, there is an additional mid-lip pressure gradient $\Delta p'_{23}$ for bilayer coatings. By a substitution of the bottom layer film height via (3.25) into (3.14), the mid-lip pressure gradient for $h_B < h_G/2$ is expressed as:

$$\Delta p'_{23} = l_M \cdot \kappa \cdot \left(\frac{(\varepsilon+1)(2\varepsilon+1)}{\varepsilon}\right)^{\varepsilon} \cdot \frac{\left(2 \cdot \frac{h}{(1+M)} - h_G\right)^{\varepsilon}}{{h_G}^{(2\varepsilon+1)}} \cdot {u_W}^{\varepsilon} \tag{3.33}$$

For $h_B > h_G/2$, the equation can be derived similarly via (3.25) and (3.15).

The combination of (3.33) with equations (3.18)-(3.21) describes the four limiting cases for power-law bilayer coatings, with $h > h_G/2$ and $h_B < h_G/2$:

Case 1 $$\Delta p'_{DU} = 1.34 \left(\kappa \frac{u_W^{\varepsilon}}{\sigma \cdot h_G^{\varepsilon-1}}\right)^{\frac{2}{3}} \frac{\sigma}{h} + l_D \cdot \kappa \cdot \left(\frac{(\varepsilon+1)(2\varepsilon+1)}{\varepsilon}\right)^{\varepsilon} \cdot \frac{(2 \cdot h - h_G)^{\varepsilon}}{{h_G}^{(2\varepsilon+1)}} \cdot {u_W}^{\varepsilon}$$
$$- l_M \cdot \kappa \cdot \left(\frac{(\varepsilon+1)(2\varepsilon+1)}{\varepsilon}\right)^{\varepsilon} \cdot \frac{\left(h_G - 2 \cdot \frac{h}{(1+M)}\right)^{\varepsilon}}{{h_G}^{(2\varepsilon+1)}} \cdot {u_W}^{\varepsilon}$$
$$+ l_U \cdot \kappa \cdot \left(\frac{(\varepsilon+1)(2\varepsilon+1)}{\varepsilon}\right)^{\varepsilon} h_G^{-(\varepsilon+1)} \cdot {u_W}^{\varepsilon} + \frac{2\sigma}{h_G} \tag{3.34}$$

Case 2 $$\Delta p'_{DU} = 1.34 \left(\kappa \frac{u_W^{\varepsilon}}{\sigma \cdot h_G^{\varepsilon-1}}\right)^{\frac{2}{3}} \frac{\sigma}{h} + l_D \cdot \kappa \cdot \left(\frac{(\varepsilon+1)(2\varepsilon+1)}{\varepsilon}\right)^{\varepsilon} \cdot \frac{(2 \cdot h - h_G)^{\varepsilon}}{{h_G}^{(2\varepsilon+1)}} \cdot {u_W}^{\varepsilon}$$
$$- l_M \cdot \kappa \cdot \left(\frac{(\varepsilon+1)(2\varepsilon+1)}{\varepsilon}\right)^{\varepsilon} \cdot \frac{\left(h_G - 2 \cdot \frac{h}{(1+M)}\right)^{\varepsilon}}{{h_G}^{(2\varepsilon+1)}} \cdot {u_W}^{\varepsilon}$$
$$+ l_U \cdot \kappa \cdot \left(\frac{(\varepsilon+1)(2\varepsilon+1)}{\varepsilon}\right)^{\varepsilon} h_G^{-(\varepsilon+1)} \cdot {u_W}^{\varepsilon} - \frac{2\sigma}{h_G} \tag{3.35}$$

Case 3 $$\Delta p'_{DU} = 1.34\left(\kappa\frac{u_W^{\varepsilon}}{\sigma\cdot h_G^{\varepsilon-1}}\right)^{\frac{2}{3}}\frac{\sigma}{h} + l_D\cdot\kappa\cdot\left(\frac{(\varepsilon+1)(2\varepsilon+1)}{\varepsilon}\right)^{\varepsilon}\cdot\frac{(2\cdot h-h_G)^{\varepsilon}}{h_G^{(2\varepsilon+1)}}\cdot u_W^{\varepsilon}$$

$$-\,l_M\cdot\kappa\cdot\left(\frac{(\varepsilon+1)(2\varepsilon+1)}{\varepsilon}\right)^{\varepsilon}\cdot\frac{\left(h_G-2\cdot\frac{h}{(1+M)}\right)^{\varepsilon}}{h_G^{(2\varepsilon+1)}}\cdot u_W^{\varepsilon}+\frac{2\sigma}{h_G} \tag{3.36}$$

Case 4 $$\Delta p'_{DU} = 1.34\left(\kappa\frac{u_W^{\varepsilon}}{\sigma\cdot h_G^{\varepsilon-1}}\right)^{\frac{2}{3}}\frac{\sigma}{h} + l_D\cdot\kappa\cdot\left(\frac{(\varepsilon+1)(2\varepsilon+1)}{\varepsilon}\right)^{\varepsilon}\cdot\frac{(2\cdot h-h_G)^{\varepsilon}}{h_G^{(2\varepsilon+1)}}\cdot u_W^{\varepsilon}$$

$$-\,l_M\cdot\kappa\cdot\left(\frac{(\varepsilon+1)(2\varepsilon+1)}{\varepsilon}\right)^{\varepsilon}\cdot\frac{\left(h_G-2\cdot\frac{h}{(1+M)}\right)^{\varepsilon}}{h_G^{(2\varepsilon+1)}}\cdot u_W^{\varepsilon}-\frac{2\sigma}{h_G} \tag{3.37}$$

In Figure 3.13 and Figure 3.14, the process windows for equations (3.29)-(3.32) and (3.34)-(3.37) are plotted for slot die 2 with a feed ratio of $M = 1$.

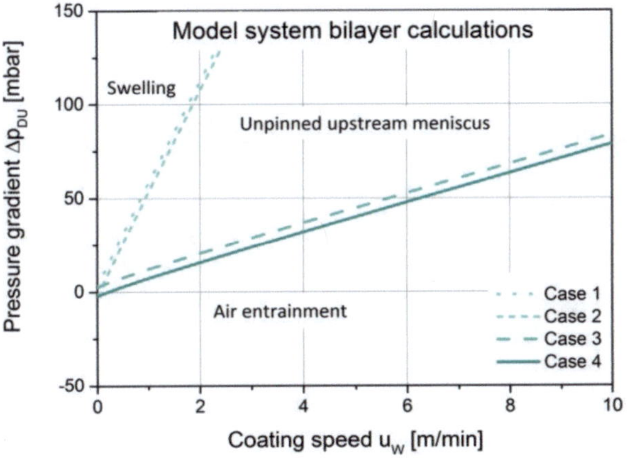

Figure 3.13: *The calculated bilayer process window for the Newtonian model system at a wet film thickness of $h = 80\ \mu m$, a feed ratio of $M = 1$ and a gap of $h_G = 148\ \mu m$.*

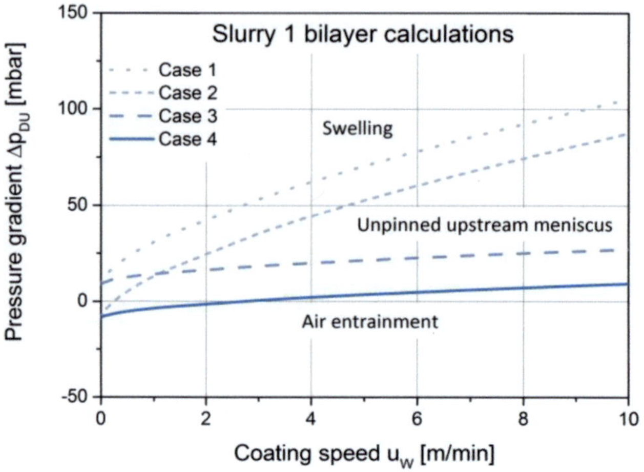

Figure 3.14: *The calculated bilayer process window for Slurry 1 at a wet film thickness of $h = 80\ \mu m$, a feed ratio of $M = 1$ and a gap of $h_G = 148\ \mu m$.*

In the above graphs, the same coating windows as in Figure 3.2 and Figure 3.3 are plotted for a bilayer application. Here, the film is equally split into top and bottom layers. As for the single-layer calculations, the power-law results are more curved in the low-speed region. The most eye-catching difference between single and bilayer coating windows are the orientations of the pairs of limits. These are now turned towards positive pressure gradients. For single-layer coatings, this trend was only observed towards thinner films, as shown in Figure 3.6 and Figure 3.7.

This difference is due to the amount of $\Delta p'_{23}$ in equations (3.29)-(3.32) and (3.34)-(3.37). For bilayer coatings with $M = 1$ and the given conditions, $\Delta p'_{12}$ and the additional pressure gradient $\Delta p'_{23}$ differ in sign in equation (3.28). Now, $\Delta p'_{12}$ must eliminate both $\Delta p'_{D1}$ and $\Delta p'_{23}$. Compared to the single-layer calculations, the onset of air entrainment starts at lower coating speeds or thicker films for $M = 1$. Thus, the feed ratio M plays an important role in bilayer coating windows.

In order to investigate the impact of M on the pressure balance (3.28), at the coating bead pressure distributions are of interest. Figure 3.15 and Figure 3.16 compare the Newtonian coating gap pressure distributions for slot die 1 and 2. For the bilayer scenario, this includes several feed ratios.

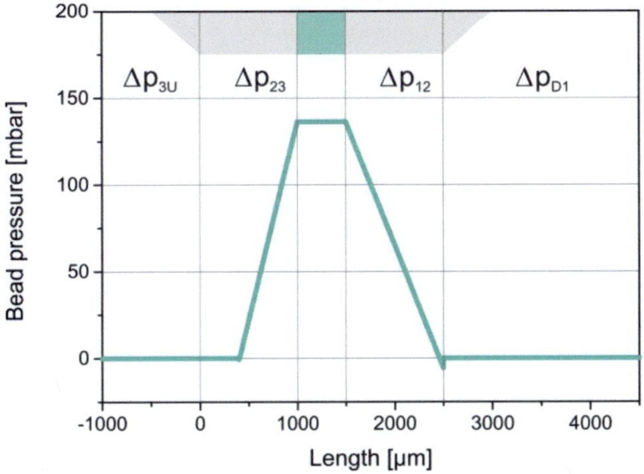

Figure 3.15: *Liquid pressure distribution of the Newtonian model system within the coating gap of slot die 1 at* $h = 120\ \mu m$, $u_W = 5\ m/min$ *and* $h_G = 148\ \mu m$.

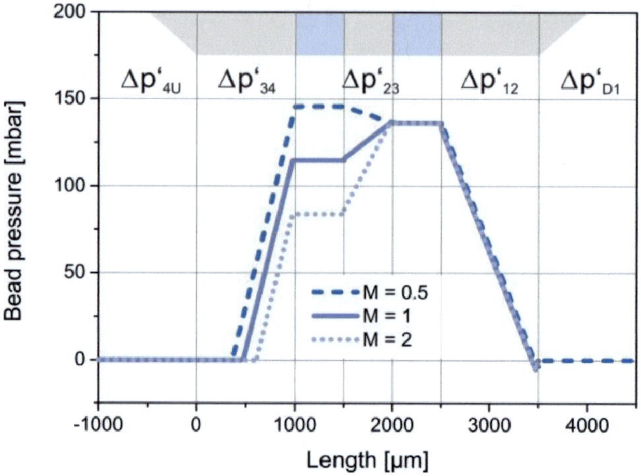

Figure 3.16: *Liquid pressure distribution of the Newtonian model system within the coating gap of slot die 2 at* $h = 120\ \mu m$*,* $u_W = 5\ m/min$ *and* $h_G = 148\ \mu m$*.*

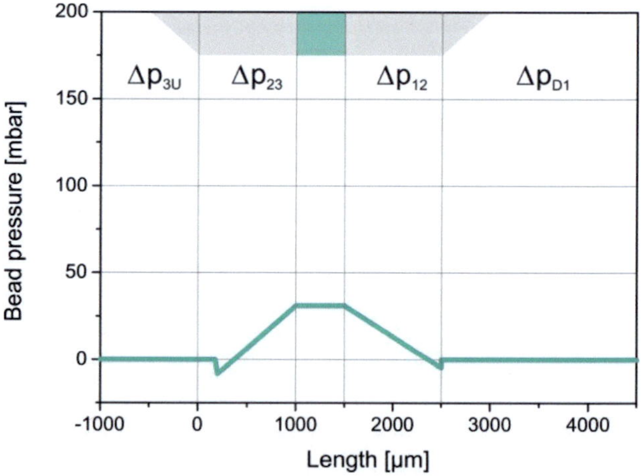

Figure 3.17: *Liquid pressure distribution of slurry 1 within the coating gap of slot die 1 at* $h = 120\ \mu m$*,* $u_W = 5\ m/min$ *and* $h_G = 148\ \mu m$*.*

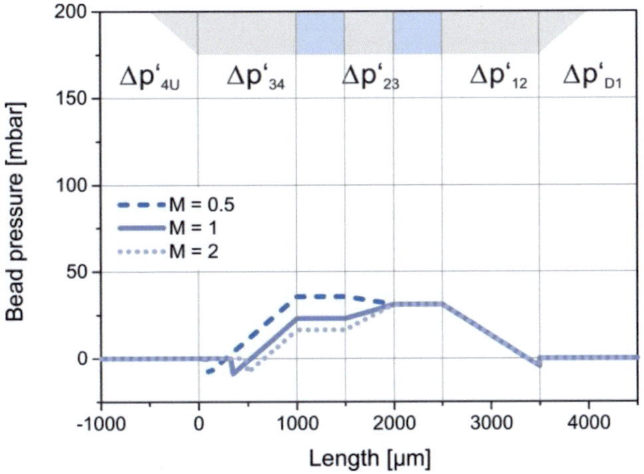

Figure 3.18: *Liquid pressure distribution of slurry 1 within the coating gap of slot die 2 at* $h = 120\ \mu m$, $u_W = 5\ m/min$ *and* $h_G = 148\ \mu m$.

The pressure distributions in the schematically drawn coating gaps plotted in Figure 3.15 to Figure 3.18 represent the conditions of $h = 120\ \mu m$ and $h_G = 148\ \mu m$. For the resulting pressure level, the upstream gap is partly wetted. The upstream meniscus has to be located at the intersection of the falling upstream pressure gradient and the ambient pressure plus or minus the capillary meniscus pressure jump. This pressure jump might react hysteresis-like for increasing or decreasing coating speeds or film heights. In the above charts, the capillary pressure jump was therefore applied positive since the goal of this chapter is to go towards the coating limits. Compared to the above conditions of $h = 80\ \mu m$, $h = 120\ \mu m$ enables a more vivid illustration of the impact of M. In general, the calculations with slurry 1 result in a lower pressure level than with the model system. This is due to the decreased viscosity of slurry 1 for these conditions ($\eta = 0.47\ Pa\ s$).

In comparison to the single layer pressure distribution, the bilayer gradients $\Delta p'_{D1}$, $\Delta p'_{12}$, $\Delta p'_{4U}$ are not changed, nor is the slope of $\Delta p'_{34}$ compared to Δp_{23}. Only the slope of the mid-lip gradient $\Delta p'_{23}$ shifts the pressure level below the bottom-layer feed slot. Thus, the starting point of $\Delta p'_{34}$, which represents the upstream meniscus location, moves along the gap.

The bead pressure conditions of single-layer coatings can be preserved for the bilayer case for $\Delta p'_{23} = 0\ mbar$. Independent of gap length, speed or viscosity, this can be achieved by choosing a critical feed ratio M_{Crit}. At the given ratio G, a sign reversal in $\Delta p'_{23}$ is reached at $M = 0.62$, which is shown in equation (3.38).

$$M_{Crit} = \frac{2h - h_G}{h_G} = 0.62 \tag{3.38}$$

If $M \rightarrow 0$, which corresponds to a dominant bottom layer, the slope of $\Delta p'_{23}$ approaches $\Delta p'_{12}$. The pressure level beneath the bottom-layer slot is then doubled, compared to the top-layer slot. In the above case, the upstream meniscus then moves outside the gap and liquid swelling at the upstream side occurs. If $M \rightarrow \infty$, the slope of $\Delta p'_{23}$ approaches the slope of $\Delta p'_{34}$, and the upstream meniscus moves towards the feed slot. Thus, case 4, representing air entrainment, could appear earlier for bilayer coatings than for single layer coatings, if $M > M_{crit}$.

Nevertheless, if one chooses $M < M_{crit}$, a bilayer and a single layer application could have the same coating window, thus leading to smaller possible film heights before air entrainment occurs. This results from a higher bead pressure beneath the upstream feed slot. For single layer coatings, this result might also be reached by an extended downstream gap.

The further implicit calculation of the onset of air entrainment (case 4) for bilayer coatings follows as presented in chapter 3.1.1.

3.2.2 Experimental results and discussion

At ambient conditions, the onset of air entrainment for Newtonian liquids in equation (3.32) is given by $\Delta p'_{DU} = 0\ mbar$. As in chapter 3.1.2, the minimum film thickness is computed as a function of the speed. To investigate its impact, the feed ratio was varied with $M = 0.1$, $M = 0.875$ and $M = 5$. To validate the calculations, the data was compared to experimental results, as shown in Figure 3.19.

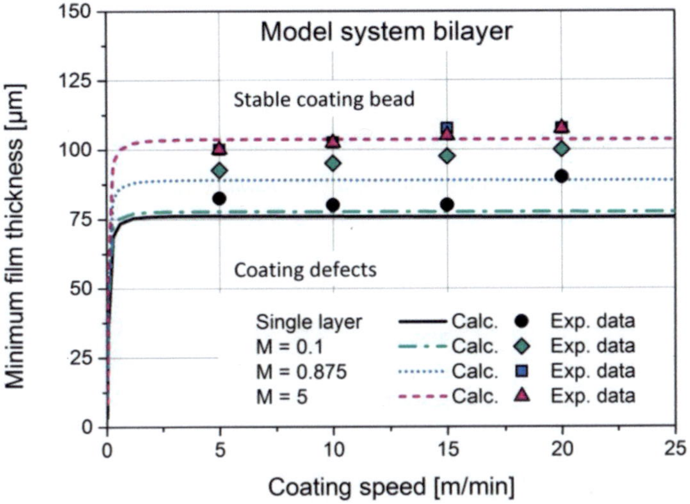

Figure 3.19: *Calculated and experimentally observed minimum film thickness of the Newtonian model system at $h_G = 148\ \mu m$ for single- and bilayer coatings with various feed ratios published in Schmitt et al. (2015b).*

The calculated and experimental results both confirm the expected impact of the feed ratio M. For the bilayer calculations, the minimum film thickness decreases from $\sim 103\ \mu m$ for $M = 5$ to $\sim 77\ \mu m$ for $M = 0.1$. The latter value approaches the single layer calculation with only $2\ \mu m$ deviation.

In contrast to the calculations, the impact of M decreases for the experimental results. While there is no observable difference between the minimum thicknesses for $M = 0.875$ and $M = 5$, they are smaller for $M = 0.1$. Still, they are at least $10\,\mu m$ thicker than for the single layer experiments. These deviations may be caused by inaccurate adjustments that decrease the impact of M, such as the gap beneath the mid-lip.

Unlike the calculations, the experiments also show a slightly rising trend in minimum film thickness towards higher coating speeds. This may be related to the simplifying assumptions, which, for example, neglect a speed dependency of $\Delta p'_{4U}$ and Δp_{3U}.

Nevertheless, the particular pressure profile of bilayer coatings can be confirmed as the cause of the increased minimum film thickness—a mechanism which can be controlled by the feed ratio M.

Based on the results presented here, another multilayer singularity, the mid-gap invasion, presented by Nam and Carvalho (2009), can be excluded as the limiting mechanism. For $h_B < 1/3\, h_G$, the theory predicts invasive, 3-dimensional flows in the upstream direction beneath the mid-lip (Nam and Carvalho, 2009). Indeed, for $M = 0.1$ and 0.875, the film defects occurred before the onset thickness was reached. Conversely, for $M = 5$, the theory predicts a mid-gap invasion with a theoretical onset film thickness of $h \approx 300\,\mu m$. However, the results do not differ significantly from those of $M = 0.875$.

The observed influence of M is also compatible with the results presented in the literature (Yu, 1995). Although this ratio was not reported to shift the minimum film thickness in any direction, it was only varied between $M = 1$ and $M = 3$. Nor do the above results differ in the range of $M = 0.875$ and $M = 5$

For slurry 1, following the power-law approach, case 4 (equation (3.37)) was again calculated implicitly. The minimum film thicknesses are plotted for several speeds in Figure 3.20, together with the single layer calculations and previous experimental results.

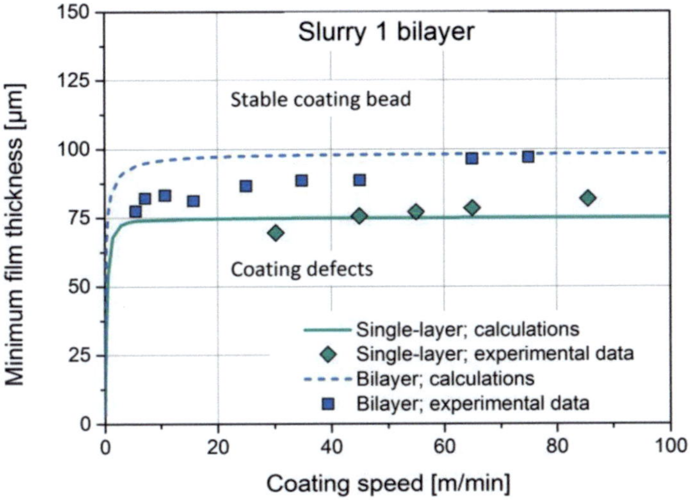

Figure 3.20: *Calculated and observed minimum film thickness of slurry 1 at $h_G = 148\ \mu m$ for single- and bilayer coatings with $M = 1$ (Schmitt et al., 2012a; Schneider, 2012; Schmitt et al., 2013b; Schmitt et al., 2015b).*

For both single-layer and bilayer applications, the minimal film thickness reaches its horizontal gradient at speeds below $0.5\ m/min$. Again, this is due to the higher value of the speed-dependent pressure gradients at lower speeds relative to the capillary upstream pressure.

Furthermore, the difference in the minimum calculated film thickness for single layer and bilayer applications is higher than for the Newtonian calculations. The value for single layer coatings remains at $h_{Min} \approx 75\ \mu m$, whereas the bilayer limit increases, relative to the Newtonian calculations, to $h_{Minimum} \approx 100\ \mu m$. This can also be attributed to the relatively higher values of the pressure gradients close to the limit. Nevertheless, if the film thickness increases, the gradients get higher for the Newtonian calculation, due to the recessive exponent in the power-law equations.

As with the calculations, the experimental results plotted in Figure 3.20 show diverging minimum film thicknesses between single layer and bilayer coatings. Here, the value of the difference matches the calculated difference quite well. In contrast, both experimental series indicate a clear rising trend for the limiting film thickness at higher coating speeds. This trend was also visible for the Newtonian results presented in Figure 3.19. As described above, the calculated values remain constant; apparently, either mechanical disturbances provoke an earlier film break-up (as mentioned in chapter 3.1.2) or a factor influenced by coating speed was underestimated or neglected in the calculations.

Pressure measurements for power-law liquids published in Schmitt et al. (2015a), confirm accuracy of the downstream pressure gradient Δp_{12} in equations (3.14) and (3.15). This might imply that the designated deviation be caused by the simplified capillary pressures. As mentioned above, Δp_{D1} was derived for Newtonian liquids and may not be suitable for describing a power-law flow. Assigning a less recessive exponent to the capillary number, and thus the speed, results in an increasing trend to the minimal film thickness. For $\Delta p'_{4U}$, there is also an expression available that includes the dynamic contact angle. Certainly, this dynamic upstream capillary pressure does not change the calculated minimum film thickness in an appreciable amount (see appendix 8.8).

Nevertheless, the observation of differing minimal film thicknesses for single layer and bilayer non-Newtonian (power-law) coatings can be explained with the model presented here.

4 Edge formation

The mechanisms creating super-elevations at the slot die coating edges remain unknown. Since they create problems in the downstream process steps, these heavy edges have to be avoided if possible. To date, the literature only contains a model for super-elevations in hot melt film-extrusion (Dobroth and Erwin, 1986). Due to its unpinned free liquid boundaries, however, extrusion coating is hardly comparable to coating onto a fixed surface.

Therefore, the available model with deformation balances should be extended to the gap region of slot die coating in the following. In the result, there should be an expression that describes the height of super-elevated edges. To validate the theories and the model, the gap ratio and the coating speed were varied in several experimental series.

4.1 Basic assumptions of the model

To obtain information about edge stresses, one must make several assumptions and take a closer look into the film-forming coating bead between slot die and substrate.

In slot die coating, the bead is bounded by four menisci. The upstream meniscus wets the substrate; the lateral menisci limit the wetting line in the cross-machine direction; and the downstream meniscus bridges the gap between die lip and film, as shown in Figure 4.1.

The position of the upstream meniscus beneath the upstream lip defines the coating process window (see chapter 1.2.1). An overstretched meniscus cannot wet the substrate properly and the resulting air entrainment limits process stability. The upstream meniscus, however, has no direct relationship to edge formation.

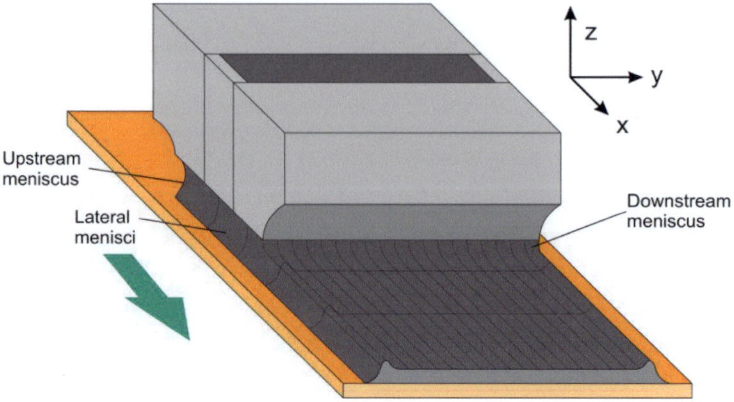

Figure 4.1: Schematic drawing of a slot die gap region with four menisci bounding the coating bead.

The two lateral menisci, along with their transition region to the upstream meniscus, affect the width of the coated film. They are formed by the contact angles of the liquid-substrate and liquid-die interface and the coating bead pressure level. Improved wettability and higher bead pressure obviously increase the coated or (at least) wetted width.

Because the downstream meniscus bridges the distance between the die lip and wet film thickness, it constantly generates and forms a new liquid surface. Here, the coating liquid gets accelerated from its specific gap flow onto the web-speed plug flow within the meniscus length. Especially in the upper part of the gap, there is a tremendous jump in liquid velocity. This may provoke a neck-in in the upper film, while the wetting line to the substrate is fixed by adhesion forces. This mechanism might be crucial for the formation of heavy edges. Thus, the downstream meniscus is attributed a high impact on heavy edge formation.

As already mentioned, one assumption is that a characteristic superposed Couette-like flow $u_G(z)$ inside the gap and a rigid plug flow u_F in the cast film which equals the web speed u_W. In Figure 4.2 the predominant flow profiles in the gap region are plotted schematically.

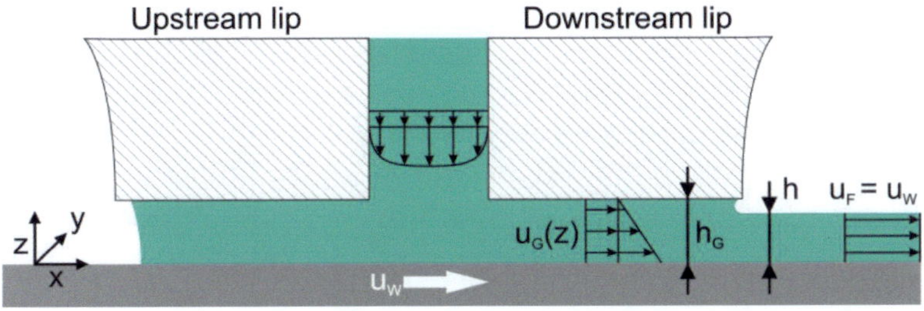

Figure 4.2: Schematic drawing of the predominant flow profiles in the slot die coating bead.

The acceleration within the downstream meniscus is the result of two mechanisms. For $G > 1$, the liquid must be accelerated from an average velocity lower than the web speed. Due to the superposed Couette flow, however, the upper liquid volume-elements are slower than average, even for $G = 1$. Since the latter mechanism depends on the flow profile, it is related to the gap ratio G and the liquid viscosity η. In combination, these accelerations provoke uneven stresses in the y-direction in the upper (not pinned) part of the cast film which, in turn, causes a neck-in and heavy edges.

At the substrate interface, the liquid is subject to the no-slip condition and does not move its wetting line or change the overall coated width. If the acceleration increases, e.g., through an increased gap height at constant film heights, the neck-in in the upper film also increases. Hence, the heavy edges should move towards the center of the film and increase in height and width, as shown schematically in Figure 4.3. Technically speaking, an increased gap heigt would lower the coating bead pressure as seen in chapter 1, thus the lateral wetting line might slightly move towards the coating center.

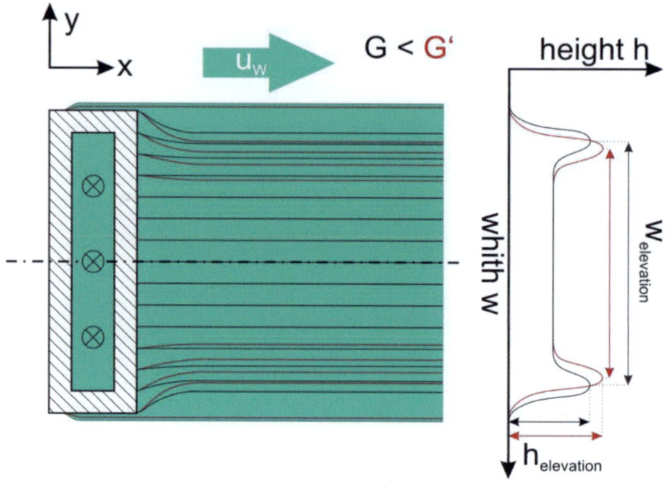

Figure 4.3: Schematic 2-dimensional drawing of streamlines in the downstream meniscus during the formation of heavy edges and the resulting film height cross-profile for different gap heights.

4.2 Existing model derivation and new approach for slot die coating

In order to develop a model of heavy edge formation, a correlation between process parameters and edge formation is needed. In analogy to Dobroth and Erwin (1986), the liquid stress situations must be balanced, starting with a volume-element. The deformation ratio ϵ_i of a liquid volume element is defined as:

$$\epsilon_i = \frac{l_i}{l_{i,0}} \tag{4.1}$$

Moreover, the product of the deformation ratios in all directions must equal unity for every in-compressible liquid volume, that is,

$$\epsilon_x \cdot \epsilon_y \cdot \epsilon_z = 1 \tag{4.2}$$

In the x or length direction, the deformation ratio ϵ_x represents the acceleration from lower gap velocity to web speed. To reduce the mathematical effort, we assume an average liquid gap velocity $\overline{u_G(z)}$. The velocity is then constant in the z or height direction. This results in an error. The generally lower and also uneven liquid flow-profile in the gap could induce super-elevations, as mentioned above. The average deformation ratio in the x direction can be calculated from the web speed and the average gap velocity as follows:

$$\overline{\epsilon}_x = \frac{u_W}{\overline{u_G(z)}} \tag{4.3}$$

For the deformation ratios in width and height, one must distinguish between the edge and the bulk region of the film. The deformation in width in the bulk is related to the averaged neck-in of the film:

$$\epsilon_{y,Bulk} = \frac{w_F}{w_G} \tag{4.4}$$

Notwithstanding the fixed wetting line at the substrate, a neck-in of the upper film decreases the averaged (over the z-direction) width w_F. Since the relative neck-in is small at slot die coating and even decreases for wider coatings it will be further neglected:

$$\frac{w_F}{w_G} = 1 \tag{4.5}$$

One of the main differences to the derivation of Dobroth and Erwin is the existence of frictional stresses in the y-direction due to the adhesive forces at the liquid-solid-interface. Since the focus of this work is on more viscous liquids, we neglect surface tensions, whereas the tensions in the z-direction are reduced to zero. This leads to the approach:

$$\sigma_y = \sigma_{Adhesion} = \sigma_z + \sigma_{Adhesion}\, for\ \sigma_z = 0 \tag{4.6}$$

Following this, the deformation in the y-direction is unequal to those in the z-direction. Introducing a new dimensionless parameter λ to describe the adhesive forces leads to the new expression:

$$\epsilon_{y,Edge} = \epsilon_{z,Edge} + \lambda \tag{4.7}$$

The parameter λ must be a function of the liquid and substrate properties and might be also dependent to process parameters, such as the gap ratio G.

The deformation ratio in the z-direction in the center of the coating corresponds to the ratio of gap height h_G to film height h. Its reciprocal is denoted and introduced here as the dimensionless gap ratio G

$$\epsilon_{z,Bulk} = \frac{h}{h_G} = \frac{1}{G}, \tag{4.8}$$

while the deformation rate in the z direction at the coating edge is defined as:

$$\epsilon_{z,Edge} = \frac{h_{Edge}}{h_G} \tag{4.9}$$

For the bulk of the film, the combination of equations (4.2), (4.3), (4.4), (4.5) and (4.8) delivers the expression:

$$\frac{u_W}{\overline{\overline{u_G(z)}}} \cdot 1 \cdot \frac{1}{G} = 1\,, \tag{4.10}$$

which can be transformed into an expression describing the unknown speed ratio by the gap ratio:

$$\frac{u_W}{\overline{\overline{u_G(z)}}} = G \tag{4.11}$$

The combination of equations (4.2), (4.3), (4.7) and (4.9) then delivers the deformation balance at the edge of the coating:

$$\frac{u_W}{u_G(z)} \cdot \left(\lambda + \frac{h_{Edge}}{h_G}\right) \cdot \frac{h_{Edge}}{h_G} = 1 \tag{4.12}$$

The substitution of the unknown velocity ratio in (4.12) by (4.11) and (4.9) leads to the quadratic equation:

$$\epsilon_{z,Edge}{}^2 + \lambda \epsilon_{z,Edge} - \frac{1}{G} = 0\,, \tag{4.13}$$

One of the two solutions of this quadratic equation (4.13) is given by:

$$\epsilon_{z,Edge} = \frac{-\lambda + \sqrt{\lambda^2 + 4\frac{1}{G}}}{2} \tag{4.14}$$

The second solution of the quadratic equation with a negative square root, which delivers negative elongation ratios, makes physically no sense and can be neglected. An expression can be derived and developed describing the height of super-elevations as a function of the gap ratio, the gap height and an unknown adhesion parameter λ which can be adapted to experimental results. The adjusted web speed is not expected to affect the edge height and the gap ratio G remains and the only fit parameter will be the parameter λ.

For a better intuitive representation of the experimental and modelling results it is reasonable to inversely substitude $\epsilon_{z,Edge}$ by:

$$E = \frac{1}{\epsilon_{z,Edge}} = \frac{h_G}{h_{Edge}} \tag{4.15}$$

With the edge ratio E drawn for various gap ratios G, the target case would be the bisecting line. Here, $E = G$ and, thus, there would be no super-elevated edges, due to $h_{Edge} = h$.

With the assumption of $\lambda = \lambda(G)$ and the substitution (4.15), the final edge model is given by:

$$E = \frac{2}{-\lambda(G) + \sqrt{\lambda(G)^2 + 4\frac{1}{G}}} \tag{4.16}$$

To validate the presented model theoretically, it can applied to the example of film-extrusion. Here, the unknown adhesion parameter has to be set $\lambda = 0$, since a film neck-in is not hindered by the contact to a substrate. The edge ratio E would then equal $\sqrt{G}$ which corresponds to the solution of Doborth and Erwin (1986) and confirms this derivation.

Within the presented model the case of $\lambda = 0$ marks the largest possible edge heights, where as the above discussed bisection of edge to gap ratio marks the target case. In Figure 4.4 these limits are plotted graphically.

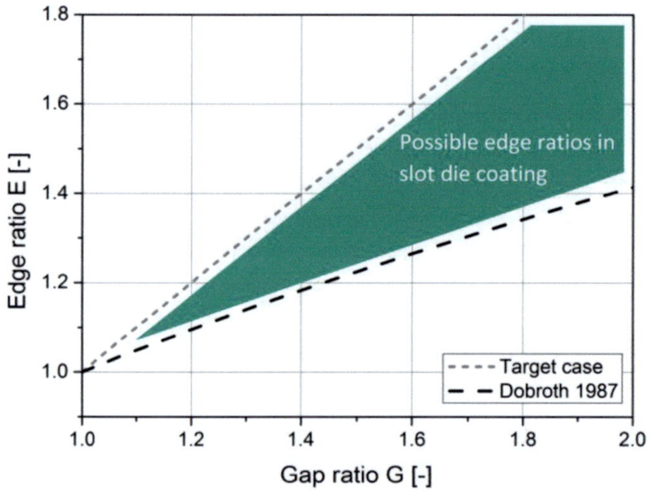

Figure 4.4: The possible edge ratios in slot die coating have to be located between the limiting cases of film-extrusion, described by Dobroth and Erwin (1986) and the target case, where edge heigt equals bulk height.

4.3 Experimental

To validate the model experimentally, slot die 3 was applied to the available experimental set-up, as explained in chapter 2.3.

The presented 2-dimensional laser triangulation system allowed to measure a neck-in and to control the steady state of the coating. The latter was reached when the actual film cross-sectional area matched a preset target value.

Since the above developed model implies an effect of the gap ratio, the gap and film heights were varied first. Three different gap heights were used: 128, 101 and $76\,\mu m$. For each of them, the film height was reduced in $10\,\mu m$ increments from an initial value close to the gap height until coating defects appeared.

In a second test series, the coating speed was varied to verify its independence from the edge height. Here, the wet film thickness was held constant ($h = 90\,\mu m$ for $h_G = 128\ \mu m$ and $101\,\mu m$; $h = 70\,\mu m$ for $h_G = 76\,\mu m$), while the coating speed was varied between $u_W = 5\,m/min$ and $20\,m/min$. In Table 4.1 the power-law approached viscosity data of the applied slurry 2 is listed.

Table 4.1: Power-law parameters and surface tensions of the applied slurry 2.

κ	66.115	[Pa s$^{\varepsilon}$]
ε	0.396	[-]
σ	0.066	[N/m]

Figure 4.5 shows three representative recordings for a gap height $h_G = 128\,\mu m$ and adjusted film heights of $h = 120\,\mu m$, $110\,\mu m$ and $100\,\mu m$ for a coating speed of $u_W = 5\,m/min$.

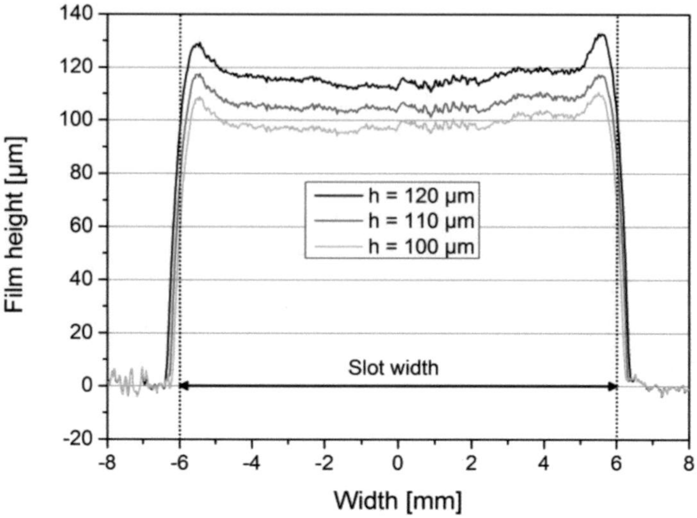

Figure 4.5: Averaged film profiles for a gap height $h_G = 128\,\mu m$ and adjusted film heights of $h = 120\,\mu m$, $110\,\mu m$ and $100\,\mu m$ coated at $u_W = 5\,m/min$. Note the reproducible marks imprinted by the relatively rough slot die lips on the film surface.

To compare the results with the model in equation (4.16), the parameter $\lambda(G)$ had to be fitted. Therefore, the method of least square sums was applied, for each experimental result. Since each fitted adhesion parameter is only valid for a particular result, a linear function was applied to the results, as seen in Figure 4.6.

The area control ratios in Figure 4.7 match quite well for the trials with $h_G = 128\,\mu m$. For the other gap adjustments, all values deviate within a range of 5%, except for one value at $G \approx 1$ and $h_G = 101\,\mu m$. Remarkably, the deviations from the target ratio only appear in the negative direction. Considering the fixed gap height and the precise syringe pump, the reason must be sought in a too high roller speed for these trials. A $0.25\,m/min$ increase in coating speed already leads to a decrease in the area ratio of 5%. For higher coating speeds, the effect of this discrete deviation is reduced by the proportional relationship of velocity to coated area. This trend is shown in Figure 4.8, where the deviation ratio approaches unity at higher coating speeds. Nevertheless, the results show a deviation in coated area of less than 5% for almost every trial, which is acceptable for an interpretation.

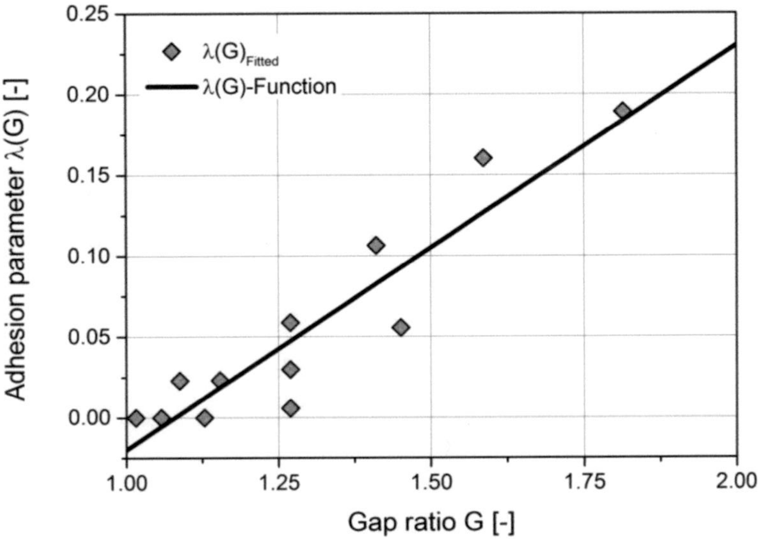

Figure 4.6: In dependence on the gap ratio G, the adhesion parameter λ was aligned by the least square sum method for the presented model to match the results. Its distribution can be described by the plotted linear function.

The calculated distribution of the correctional ratio is approximately given by:

$$\lambda(G) = 0.25\,G - 0.27 \tag{4.17}$$

Because the viscosity affects the flow profile inside the gap, this function might vary if the applied liquid system is changed. Thus, the model for E in equation (4.16) becomes:

$$E = \frac{2}{-0.25\,G + 0.27 + \sqrt{(0.25\,G - 0.27)^2 + 4\frac{1}{G}}} \tag{4.18}$$

4.4 Results and discussion

To avoid misleading interpretations of non-steady coatings (as happened in previous works (Schmitt et al., 2013c)), the coated cross-sectional area have to be controlled first. In Figure 4.7 and Figure 4.8, the ratios of preset and measured cross-sectional areas are plotted for all experimental trials. The area control ratio should match unity and is defined as:

$$YZ_{Area} = \frac{yz_{Area,measured}}{w_{Slot} \cdot h_{Preset}} \tag{4.19}$$

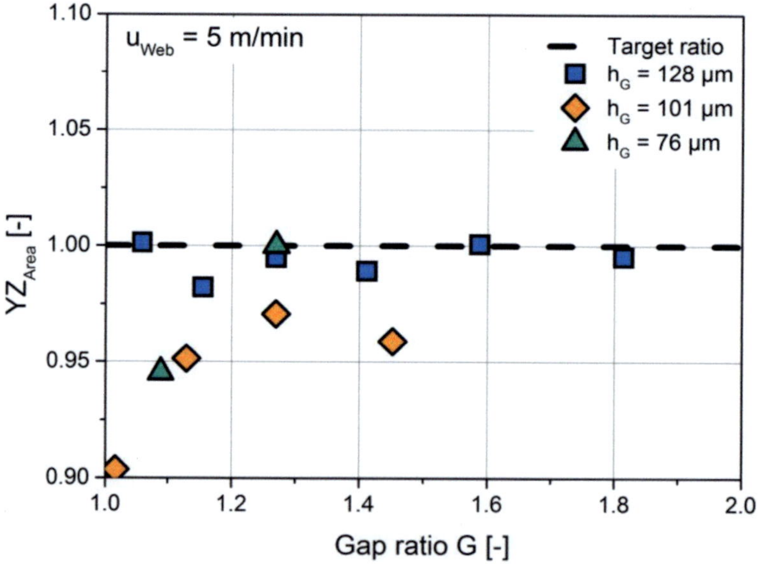

Figure 4.7: Average area ratios for the experimental series at various gap ratios and specific gap heights. The values should equal 1.

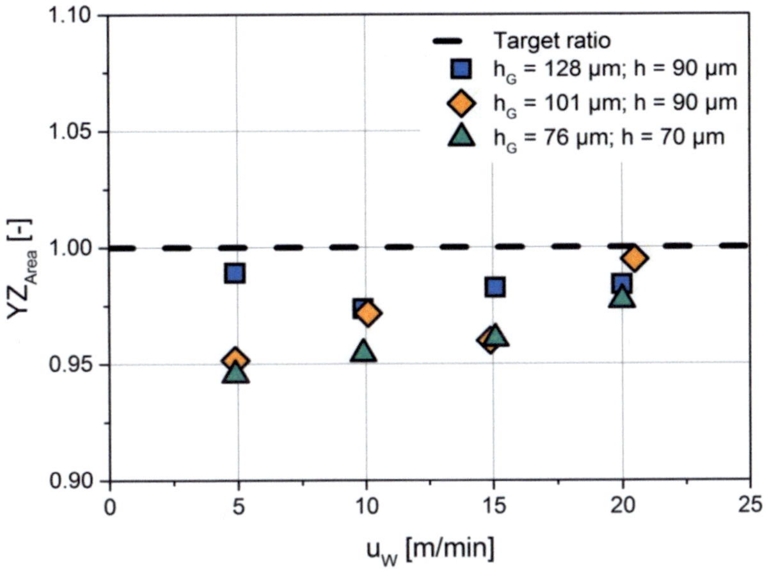

Figure 4.8: Average area ratios for the experimental series at several coating speeds and specific gap heights. The values should equal 1.

The representative profiles plotted in Figure 4.5 show three important characteristics of the experimental set-up. The first is the roughness or inhomogeneity of the film bulk. Here, the self-constructed slot die seems to shape its lip profile onto the film surface. The second characteristic is the wider wetting for increasing film heights. As mentioned above, a higher volume flow leads to higher gap pressures, which pushes the lateral menisci outwards. The last, but most important, characteristic concerns the resulting bulk film height. Because the film is relatively narrow ($w_{Slot} = 12\,mm$), a laterally extended wetting line and higher heavy edges (relatively to the film height) lower the overall film height significantly.

In Figure 4.9 and Figure 4.10, the measured edge heights are plotted for the measured bulk heights of the applied films. The deviation in edge height is the measured difference between the left and the right edges. Similar, the drawn bulk-height deviation represents the measured deviation between the left and right part of the bulk.

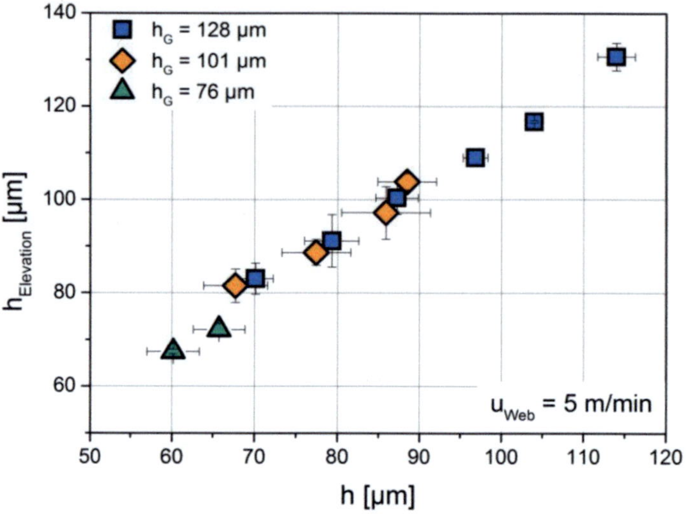

Figure 4.9: Measured average edge height for different applied film and gap heights.

The results of the measured edge heights for different preset film and gap heights show a clear trend. The edge height increases proportionally to increasing film heights. Overall, one observes a $9 - 19\%$ super-elevated edge. The general gradient of the trend is approximately the same for all gap heights. Nevertheless, the values for $h_G = 76\ \mu m$ are beneath an imagined line connecting all values of $h_G = 101\ \mu m$ and $h_G = 128\ \mu m$. This may be related to the presentation method of the data. As mentioned above, the acceleration of the liquid may be the crucial edge formation mechanism. For example, the wet film of $h \approx 70\ \mu m$ coated at three different gap heights experienced different accelerations. As expected, the edges increase with the gap height.

Another remarkable result is the confirmed characteristic of the lowered film height, as can be seen for $h_G = 128\,\mu m$. For lower expected gap pressures (see Figure 4.9, $h_G = 128\,\mu m$, $h \approx 70\,\mu m$), the measured films are closer to the preset height than for higher pressures (see Figure 4.9, $h_G = 128\,\mu m$, $h \approx 120\,\mu m$).

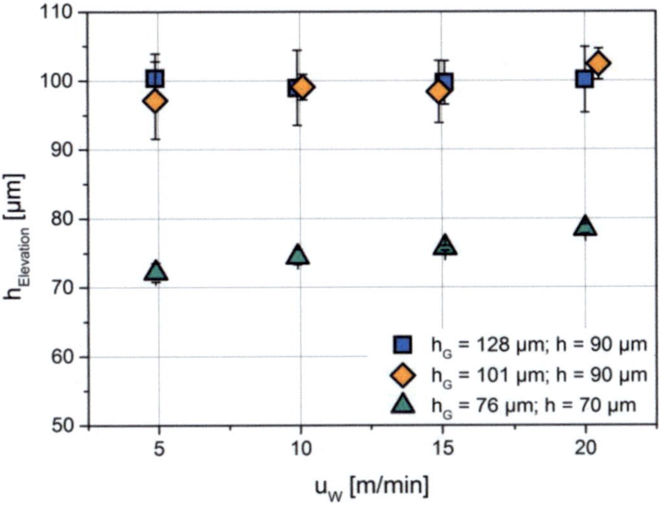

Figure 4.10: Measured average edge height for differen gap heights and coating speeds.

In Figure 4.10, the results for $h_G = 101\,\mu m$ and $h_G = 128\,\mu m$ do not show any correlation between edge height and coating speed. This confirms the above-mentioned theory. The ratio of averaged gap and coating speed does not change for higher speeds and a constant film height. Thus, the acceleration stays constant. However, the error made by averaging the gap speed should increase, due to the more extreme flow profiles for higher speeds. For the speed trials at $h_G = 76\,\mu m$, the film height was reduced from $h = 90\,\mu m$ to $70\,\mu m$. This explains the overall lower edge heights of these series. In contrast, the rising trend of the edge heights with increasing coating speeds can be explained with Figure

4.10. Obviously, this trend must be related to an increased cross-sectional area coated at higher speeds. Referring the results for different gap heights in Figure 4.9 to a common base, their edge ratio E can be plotted for the applied gap ratios G, as shown in Figure 4.11.

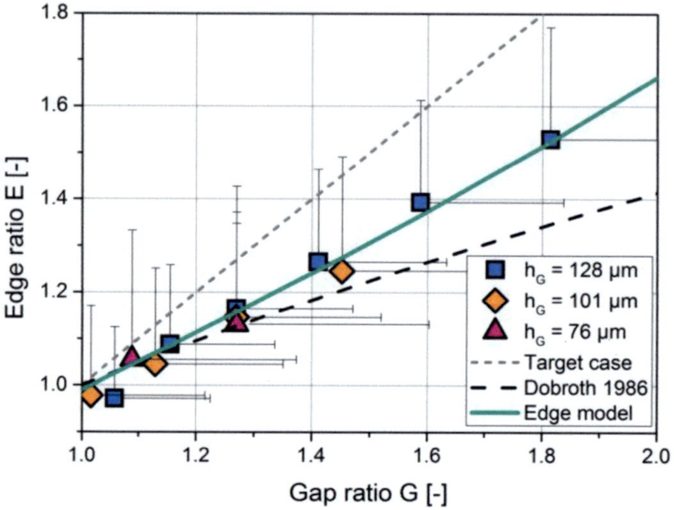

Figure 4.11: *Measured and calculated edge ratios E plotted for the applied gap ratios G. The narrow-dashed line marks the theoretical ideal state with no super-elevated edges, from which the results depart for higher gap ratios.*

Plotted as the ratio E, the results in the above Figure 4.11 line up on an increasing trend. Hence, this type of presentation seems to include all influencing parameters in the observed range. Except for coatings close to $G = 1$ all results are located in the assumed range as presented in Figure 4.4. In detail, the results of E show a general trend of departing from the ideal state for increasing G, which confirms the theory and the observations of higher edges for increasing gap ratios. The relatively wide error bars are due to the staggering roller. Due to the gap adjustment at the tightest spot, they are oriented only in positive direction. For gap ratios

G approaching unity, the observed edge ratio E even falls below unity. This results in edge heights higher than the adjusted gap.

The presented edge model in equation (4.18) describes the results quite well for gap ratios of $G > 1.1$. This was achieved by an increasing fitted parameter $\lambda(G)$ towards higher gap ratios, as presented in Figure 4.6. Since the origin of λ should be the interfacial adhesion at the film edges, this might be due to the shear depended viscosity. Towards increasing gap ratios the the slurry viscosity is also increasing. A higher liquid viscosity is counteracting the film motion leading to heavy edges. This might also explain the complete absence of super-elevations for slurries with high viscosities at low shear-stresses (Bitsch et al., 2014).

In contrast, for gap ratios towards $G = 1$, the model underestimates the height of the super-elevated edges, due to the neglected influence of the flow profile in the gap. The results show that a coating without heavy edges is not possible with the applied LIB anode slurry. Due to the adhesive parameter's dependence on the applied liquid properties, this may be improvable with higher low-shear viscosities as shown by Bitsch (2014).

Nevertheless, the edge model is capable of predicting super-elevated edges for the applied liquid system. The results further showed that heavy edges can not be prevented for slurry 2, but reduced by the knowledge of the presented model. An optimum seems to be found at $G \approx 1.1$. Below this ratio the disparity to the target case is no longer decreasing resulting in a slightly increase of the edge height, compared to the bulk height. This comes in handy for the industrial practice, where very narrow gap heihgts are difficult to adjust due to machine tolerances.

5 Intermittent slot die coating

In slot die coating there is a direct relationship between coated wet film thickness and liquid pressure distribution in the die and upstream feed system. For casting a homogeneous film intermittently, this liquid pressure has to reach a steady state with a compulsory leveling at each coating start. Although there are many patents for intermittent slot die coating, the dynamic part of this process has not yet been covered by scientific research.

The following chapters discuss the process parameters that most strongly influence intermittent coatings. A steady state model is also developed that is capable of predicting the optimum environmental conditions (Diehm, 2014; Spiegel, 2014). The model is then validated using the experimental-set up presented in chapter 2.3.4.

A final characterization of the available set-up will identify and discuss the speed limits for valve-supported intermittent coatings.

Parts of this chapter have been published previously by the author (Schmitt et al., 2015a).

5.1 Steady state model

The following chapters will discuss which process parameters affect the coated film and how they behave for an intermittent film casting. A model will then be developed to reveal the basic needs of high quality intermitted films, while compared to experimental results.

5.1.1 Assumptions

Literature showed a direct relation between liquid pressure in the coating feed system and the afterwards casted film (Kim S et al., 2009). In Figure 5.1 the dynamic interaction of film thickness h and the appropriated die pressure Δp_{Die} during an intermittent coating is plotted. The process parameters were adjusted to $u_w = 5\,m/min$, $h_G = 127\,\mu m$, $h = 120\,\mu m$, $l_C = 60\,mm$ and $l_I = 40\,mm$.

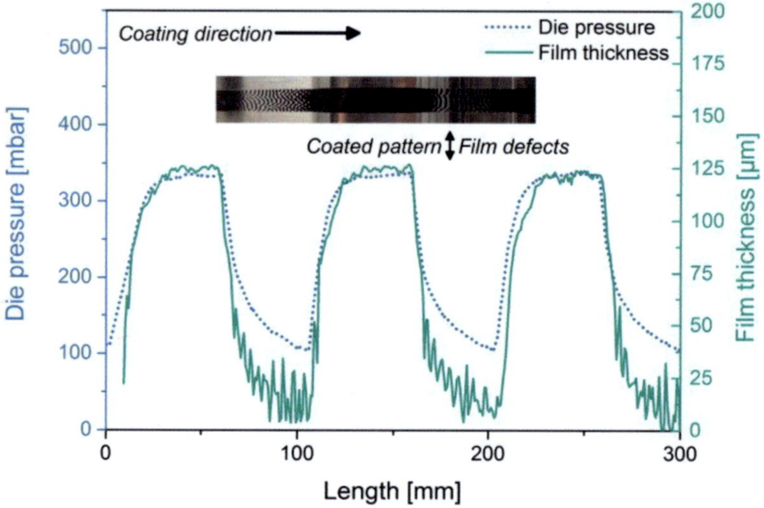

Figure 5.1: Die pressure (blue dotted line, left) and wet film thickness (green line, right) during the intermittent coating of an electrode film. The length axes were aligned to the start of decreasing trends (Schmitt et al., 2015a).

The above figure shows the periodically alternating die pressure (blue dotted line) for three consecutively coated patterns. Even though the length axis was shifted and the patterns are randomly combined with the film profile, the wet film thickness (green line) follows the pressure distribution exactly. In every coating phase, the die pressure reaches a plateau. Here, a steady state is reached in which the flow and, thus, the pressure drop in the die, is constant. The closing bypass valve at the end of the coating results in a decreasing volume flow and hence a decreasing pressure drop. Once the bypass valve is fully closed, the die pressure does not drop to zero immediately. Perhaps due to its viscoelastic behavior or other effects like entrapped compressible, microscale air bubbles, inertia or capillary flow, the electrode slurry releases its stored energy and expands. This leads to a continued but decreasing slurry flow leaking onto the substrate. The photographic image in Figure 5.1 shows that this flow is beneath the coating limit; consequently, film defects appear. Most patents mentioned in chapter 1.2.4 cover some kind of technology to avoid this slurry leaking. They provide a sharp stop edge by either drawing back the slurry, manipulating the gap, or both. The leaking cannot be avoided by the use of a turning bypass valve alone, as was was applied in this work. Therefore the avoidance of slurry leaking between the coated patterns is beyond the scope of this work. At the chosen process settings, the bypass valve was opened again before the slurry could fully expand. Hence, no steady-state die pressure in the interruption phase is observed. Opening the bypass valve again does not increase the pressure abruptly. Again, this may be attributed to the viscoelastic slurry behavior and inertial flow resistance. The nearly resting and elastic slurry first has to be accelerated. After this ramping time, the wet film thickness increases towards the preset value. Remarkably, the slope of the die pressure for the onset (i.e. opening the valve) is less steep than for the offset (i.e. closing the valve). This is attributable to the one-sided spring support of the valve.

The results observations above show the direct link between the die pressure and the deposited film. Thus, it is crucial to control this pressure for optimizing the film profile and holding back the expanding slurry. However, before optimizing the ramping or leaking via advanced intermitting technologies, one has to take a closer look at the upstream system pressure. For a minimized pattern ramping time, the pressure drop of waste pipe and slot die should be equalized. A disparity could provoke an additional leveling in the whole liquid. It might even prevent steady state for the coating and interruption phases, as seen later on. Hence, the expected system pressure during the coating phase must be calculated and the waste pipe adjusted accordingly.

5.1.2 Model derivation

To predict the steady state liquid pressure during the coating and interruption phases, one has to calculate the sum of expected pressure drops in the feed system. For a comparison between model predictions and experimental results its reasonable to start the calculation at the upper pressure transducer (see Figure 2.9).

Starting from this point, first an expression for parabolic pipe flow is needed. The pressure drop in a pipe is given by the Hagen-Poiseuille expression. Modified with the representative shear rate for a power-law fluid in a circular pipe (VDI-Gesellschaft Verfahrenstechnik und Chemieingenieurwesen and Gesellschaft, 2010), the expression becomes:

$$\Delta p_P = \frac{128 \cdot l_p}{\pi} \cdot \kappa \cdot \left[\frac{32}{\pi} \cdot \left(\frac{4}{\frac{1}{\varepsilon} + 3} \right)^{\frac{1}{1-\frac{1}{\varepsilon}}} \cdot \right]^{\varepsilon - 1} \cdot \frac{(h \cdot w_S)^{\varepsilon}}{{d_P}^{3 \cdot \varepsilon + 1}} \cdot {u_{Web}}^{\varepsilon} \tag{5.1}$$

Due to the pre-metered character of slot die coating, the flow rate could be substituted for using the die width, coating speed and film thickness (see equation (1.1)). The same approach applies for the flow inside the slot die. Combining the expression of pressure drop for a slot flow with its representative shear rate (VDI-Gesellschaft Verfahrenstechnik und Chemieingenieurwesen and Gesellschaft, 2010) yields the expression:

$$\Delta p_S = 12 \cdot h_S \cdot \kappa \cdot \left[6 \cdot \left(\frac{3}{\frac{1}{\varepsilon} + 2} \right)^{\frac{1}{1-\frac{1}{\varepsilon}}} \right]^{\varepsilon - 1} \cdot \frac{h^{\varepsilon}}{{l_S}^{2 \cdot \varepsilon + 1}} \cdot {u_W}^{\varepsilon} \tag{5.2}$$

Describing the pressure drop in the coating gap is more complex. Due to a moving wall (the moving substrate), the Hagen-Poiseuille flow is superposed by a Couette flow. Durst (Durst and Wagner, 1997) gives an expression for Newtonian fluids:

$$\Delta p_G = 1.34\, Ca^{\frac{2}{3}} \frac{\sigma}{h} + l_D \cdot \left(1 - \frac{2h}{h_G}\right)\left(\frac{6\, \eta\, u_W\, l_D}{h_G^2}\right) \tag{5.3}$$

For $2h > h_G$, Lee (2011) solves an expression for a power-law pressure drop inside a coating gap:

$$\Delta p_G = 1.34\, Ca^{\frac{2}{3}} \frac{\sigma}{h} + l_D\, \kappa \left(\frac{(\varepsilon+1)(2\varepsilon+1)}{\varepsilon}\right)^{\varepsilon} \cdot \frac{(2 \cdot h - h_G)^{\varepsilon}}{{h_G}^{(2\varepsilon+1)}} \cdot {u_W}^{\varepsilon}, \tag{5.4}$$

wherein the first summand describes the pressure beneath the downstream meniscus, known from the correlation by Ruschak (see chapter 3.1.1). Although this correlation was derived for Newtonian liquids, only minor deviations were assumed, due to a small h_G to h ratio and the overall small value of this part. The capillary number in the coating gap for power-law fluids is approximately given by equation (3.17).

The resulting expression for the system pressure between upper pressure transducer and environment consists of various pipe and slot terms as well as one gap term. Pipe bending and the transverse pressure drop inside the die cavities were neglected:

$$\Delta\, p_{System} = \sum_i \Delta p_{P,i} + \sum_j \Delta p_{S,j} + \Delta p_G \tag{5.5}$$

During the interruption phase, the liquid gets directed into a waste-pipe. Hence, the system pressure then consists only of several pipe terms. For calculating the die pressure inside the downstream die cavity, only the pressure drops of the first slot and the coating gap need to be included:

$$\Delta\, p_{Die} = \Delta p_{S,1} + \Delta P_G \tag{5.6}$$

5.1.3 Experimental results and discussion

A high impact is assumed for deviations in rheology and dimensions to the equations (5.1), (5.2) and (5.4). Therefore, the calculations should be validated for the shear thinning anode slurry 2 and for the Newtonian model system. The liquid data of the coated slurry 2 and the Newtonian model system are listed in Table 3.1. As mentioned above, each slurry batch is slightly diverging in viscosity. Therfore, the values of slurry 2 applied in chapter 1 are diverging from those listed below.

Table 5.1: Power-law parameters, viscosity and surface tenisions of the applied slurry 2 and the Newtonian model system.

	Slurry 2	Model system	
κ	59.4	-	[Pa s$^{\varepsilon}$]
ε	0.37	-	[-]
η	-	1	[Pa s]
σ	0.066	0.018	[N/m]

The validation was realized by the comparison of computed and measured liquid pressures in the feed system and the die during continuous coatings of various speeds. The results are plotted in Figure 5.2 and Figure 5.3. As described above, the gap and the film thickness were held constant. Therefore, a rising coating speed results in an increasing volume flow and, thus, an increasing pressure drop.

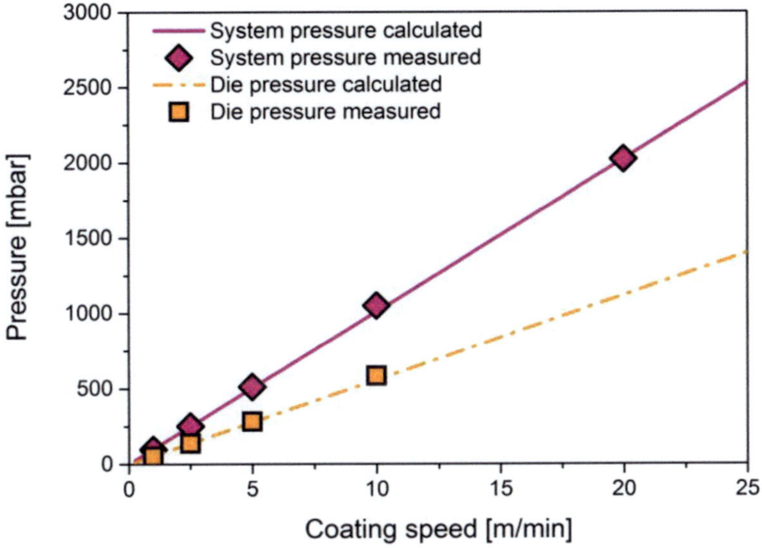

Figure 5.2: Calculated system and die pressure for the Newtonian model system compared to the measured values during coating at various speeds ($h_G = 128\ \mu m$, $h = 90\ \mu m$).

For the Newtonian model system (see Figure 5.2), the calculations match the observed pressures in die and system quite well, thus confirming the calculation model. Towards higher coating speeds, the fluid pressure increases linearly.

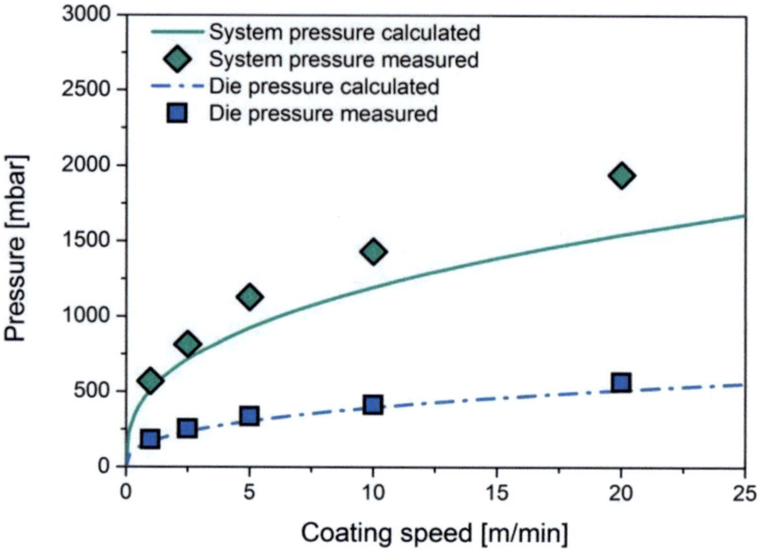

Figure 5.3: Calculated system and die pressure for slurry 2 compared to the measured values during coating at various speeds ($h_G = 128\ \mu m$, $h = 90\ \mu m$) (Schmitt et al., 2015a).

For the anode slurry (Figure 5.3) with the shear thinning fluid character, the pressure gradient decreases, due to velocity exponents smaller than unity in equations (5.1), (5.2) and (5.4). The results in the above chart confirm the model for the pressure drop of coating gap and die slot (blue dotted line and squares). The calculations match the measured die pressures for various coating speeds very well. In contrast, the pressure drop through the whole system seems to be underestimated. The measured pressure is about 10% higher than the calculated (green line and diamonds). This may be related to the neglected curves or unknown constrictions between the die and the second pressure transducer.

In the next step, the model was used to predict the steady-state pressure during the coating phase of the intermittent process. To show the impact of mismatching pressure drops on the intermittent process, three different waste pipe lengths were mounted. The first one was designed to match the calculated pressure drop in the die direction, the second one had only 70% ($\Delta p_{WP} < \Delta p_C$) of the predicted pressure drop and the third one, 150% ($\Delta p_{WP} > \Delta p_C$) of the desired value. In Figure 5.4, the measured system pressures and the resulting film profiles are plotted.

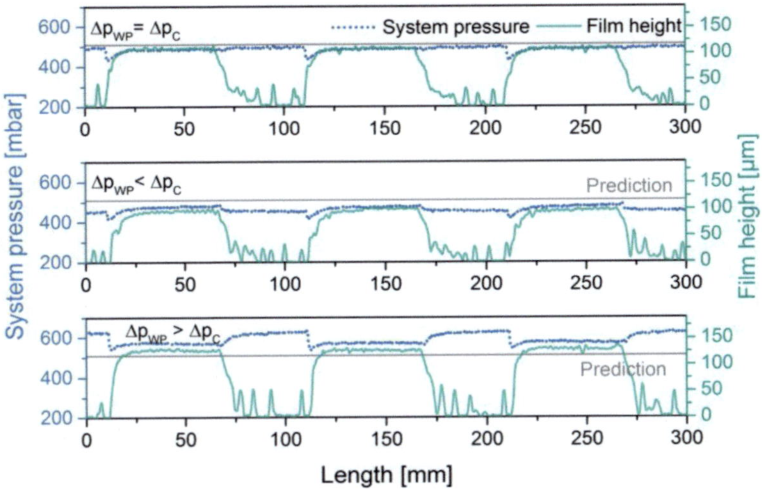

Figure 5.4: System pressure, its prediction and the related film thickness for standard intermittent coatings at $u_W = 1\,m/min$ plotted for three different waste pipe pressure drops. The variation was realized by means of different bypass-pipe lengths. Without a pressure alignment, too thin films or a die swell appear for a too high or low bypass pressure level (Schmitt et al., 2015a).

The best coating is expected for equal pressure drops in slot die and waste pipe system (Figure 5.4; *top*). The results show a constant pressure during the whole coating process, only interrupted slightly during the valve switching step. For this adjustment, the measured wet film thickness reaches the predicted plateau during the coating phase. Likewise, the system pressure and the film thickness also show a ramping at each coating start and stop. During the interruption phase, the expanding slurry inside the slot die leaks onto the substrate, causing small barring lines (similar to the photographic image in Figure 5.1). To avoid the leaking additional equipment for a slurry draw back or an alternating gap would be needed (which is beyond the scope of this work). The drop in pressure after each switching may be attributable, as described above, to the viscoelastic slurry behavior and flow resisting effects, such as inertia, depending on the switching direction. The time needed to accelerate the slurry to a steady state depends on the slurry volume. According to this, it takes more time to accelerate the slurry in the coating die than in the waste pipe with a tenth of the die volume. For the coating direction, a steady coating bead also has to be developed. As described elsewhere (Chang et al., 2009), this step is faster at higher coating speeds, smaller gap volumes and higher slurry flow rates. The results underscore the influence of the bypass valve. It reduces the maximum resting slurry volume approximately to the inner volume of the slot die. Developments in industry even push the valves inside the die (Janssen, 2011; Huth and Sekler, 2014), in order to reduce the amount of leaking by minimizing the relaxing slurry volume. The observed drop in pressure during the switching strongly depends on the slurry rheology and on the set-up dimensions. Especially for large volume production dies, this effect may be more distinct. Unless a feasible correlation is established, the influence of each slurry-equipment interaction has to be measured individually. For Newtonian liquids, where only inertia, capillary effects and the developing contact line damp the flow, one would expect a shorter transition state.

In contrast to the optimum for equal pressure drops in slot die and waste, a too low waste pipe pressure drop may even prevent the coating of a homogeneous film (see Figure 5.4, *center)*. The measured system pressure alternates between the predicted coating and waste pipe pressures by slightly rising and falling. For the waste pipe lengths used here, the system does not reach any steady state. The pressure at the coating start rises too slowly to reach a stable coating bead within the first centimeters. Even after the entire coating period, the predicted film height was not reached. Thus, a much lower waste pipe pressure may prevent a closed film within the coating phase. Considering the conservation of mass, this could only be explained by a still rising trend, which is not clearly visible for this short coating period. Hence, different system pressures not only affect the quality of the coating but definitely limit the whole process.

If the waste pipe pressure drop is too high (Figure 5.4; *bottom),* the effects are reversed. The system pressure alternates but stays constantly higher than predicted during the coating. After a short pressure drop during switching to the coating phase, the pressure remains too high. Furthermore, the resulting wet film is thicker than the adjusted gap.

This effect might be explainable by the viscoelastic slurry behavior or inertial forces. The constant volume flux during the interruption places the whole liquid system under higher pressure. This causes the released slurry to excessively flow through the die and provoke a die swell at the outlet. The stored energy in the relatively large upstream pumping system seems to be enough to expand inside the whole pattern length. Again, this effect inhibits the intermittent process and emphasizes a prior pressure calculation. On the other hand, as a possible advantage, this effect could shorten the ramping at each coating start. A temporary limited expansion into the die might support the leveling at the starting edge.

Nevertheless, the pressure and the film height should show a trend towards their prediction, which is not visible in the present coating time. To prove this assumption, additional experiments with longer coating and interruption phases are necessary. Thus, in Figure 5.5, the results for different coating lengths of intermittent coatings with a too high waste pipe pressure drop are plotted.

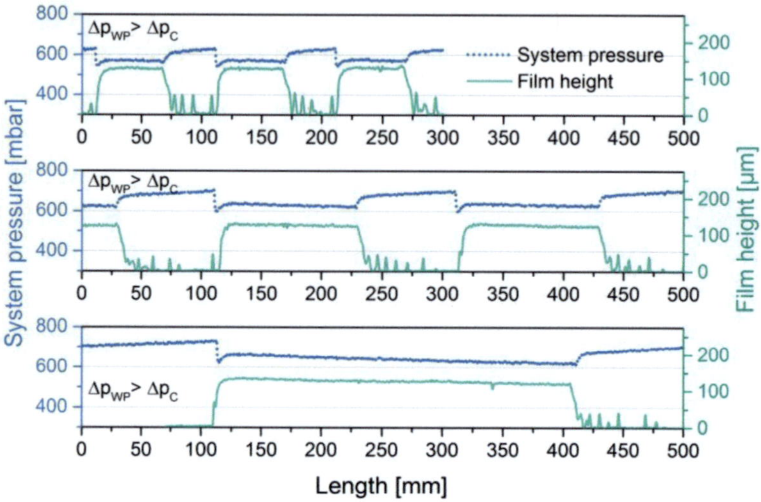

Figure 5.5: System pressure and the related film thickness for intermittent coatings at $u_W = 1\ m/min$ and a too high waste pipe pressure drop plotted for three different period lengths. The periods were varied from $60/40\ mm$, to $120/80\ mm$ and $300/200\ mm$ for the coating and interruption lengths, respectively (Schmitt et al., 2015a).

The above chart shows that the applied liquid system compensates an overly long waste pipe very slowly. The predicted trend of falling pressure and film height during the coating phase is only observable for longer coatings in the range of $l_C = 300\ mm$. Only for this long term coating does the film finally reach its predicted film height. These experiments were repeated for intermittent coatings with a too low waste pipe pressure (see appendix 8.9). The results confirmed the above observations, but in reverse, with an increasing liquid pressure during the coating.

If this long-term dampening and the short-term pressure drop during the switching is caused by viscoelasticity they should not appear for Newtonian liquids. Thus, the experiments in Figure 5.4 were redone with the Newtonian model system to first investigate the valve switching impact. The results are plotted with those of the LIB slurry in Figure 5.6.

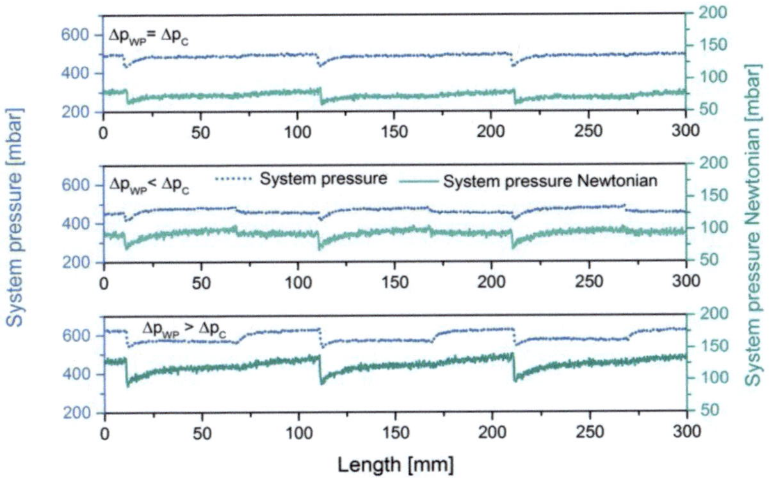

Figure 5.6: System pressure of the LIB slurry and the Newtonian model liquid for intermittent coatings at $u_W = 1\,m/min$ plotted for three different waste pipe pressure drops (Schmitt et al., 2015a).

The system pressure of the Newtonian liquid shows quite the same curvature as the LIB slurry when reacting to the valve switching. Due to the different pressure levels for each liquid, it is difficult to compare their curvature in amount or detail. Yet, inertial forces or entrained air in the slurry seem to be the underlying cause of the damped acceleration after each switching state.

To investigate the long term dampening effect on Newtonian liquids, the experiment in Figure 5.6 bottom was repeated with the model system. The results are plotted in Figure 5.7, in comparison to those of the LIB slurry.

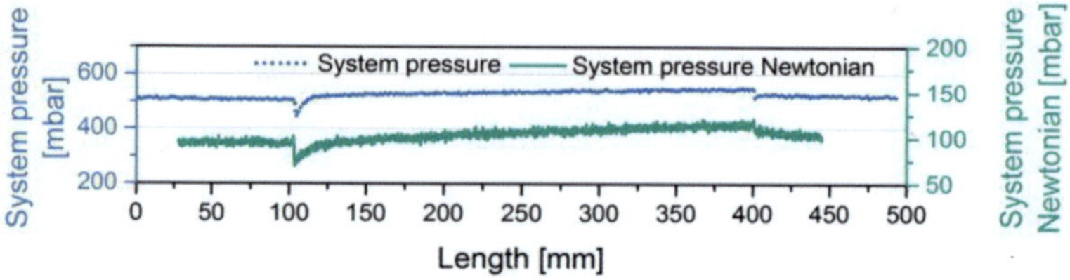

Figure 5.7: System pressure of the LIB slurry and the Newtonian model liquid for intermittent coatings at $u_W = 1\,m/min$*, a too low waste pipe pressure and a coating phase of* $l_C = 300\,mm$ *(Schmitt et al., 2015a).*

Similar to the results of the LIB slurry, the system pressure of the Newtonian liquid increases during the whole coating phase. Thus, this effect is also not caused by slurry viscoelasticity but mainly by inertial forces. Other possible dampening sources may be enclosed air bubbles trapped in undercuts or elastic walls. However, the system was carefully bled during the filling operation. Moreover, the liquid containing materials were selected to be very stiff, as described above.

In observing the too low and too high coatings in Figure 5.4, two questions arise: First, where is the not casted liquid going to and second, where is the excessive liquid coming from? The conservation of mass would seem to be violated, since the slurry is not as compressible as a gas. In order to disprove this, the waste pipe flux in each associated interruption phase has to compensate the deviation in coated height.

This assumption could be proved by the comparison of applied and predicted liquid volume for the combined coating and intermitting phases ($v_C + v_I = 0.144\,ml$). The applied volume, therefore, was calculated implicitly from the measured system pressure. The obviously different calculations of waste pipe and coating direction are separated by the switching points.

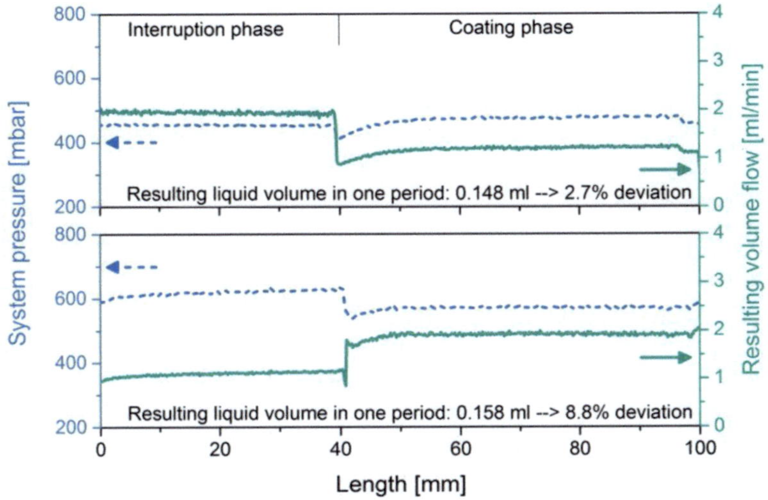

Figure 5.8: System pressure of the LIB slurry and resulting volume flow at $u_W = 1\ m/min$ for a too low (top) and a too high waste pipe pressure (bottom) (Schmitt et al., 2015a).

As predicted for a too low waste pipe pressure, the volume flow during the interruption phase is far higher than for the coating phase (Figure 5.8, top). When integrated for the whole period, the applied volume amounts $v = 0.148\ ml$, which deviates only 2.7% from the pre-set value. The observation is reversed for a too low waste pipe pressure. The volume flow during the interruption phase is much smaller than during the coating and results in an applied volume of $v = 0.158\ ml$ and a deviation of 8.8%. Even with a critical look at the calculated values, the assumption of a compensating interruption phase could be confirmed.

Concluding, a precise adjustment of the upstream pressure levels is crucial for intermittent coating with a bypass-valve. Otherwise, inertia effects imply a long-term dampening and thus affect the coating quality, in dependence to the applied system.

The short term dampening responsible for the ramping at every coating start and stop was also attributed to inertia. This ramping and the also observed slurry leaking between the coatings, might not be improvable by a bypass-valve alone. For sharp starting and stopping edges, additional techniques are necessary, such as those already filed as patents (see chapter 1.2.4) or invented by the author (KIT, 2014).

5.2 Limitations for intermittent coatings

An estimation of the final limits for intermittent coatings is nessecary regarding high speed applications. Thus, in the following theoretical borderlines will be compared to actual measured values for the above presented experimental set-up.

5.2.1 Theory

Within certain limits, an increased coating speed lowers the specific film manufacturing costs and improves production capacity. If the coating is projected to be intermitted, the presented coating process window is no longer the only speed-limiting factor. Every intermittent coating method has a minimum switching time between the states of fully established coating flow and complete interruption, both when starting and stopping flow. This time interval might be further broken down into real hardware switching and the above-mentioned liquid flow ramping. Thus, the overall switching time of a method limits coating speed for any given interruption or coating length. In Figure 5.9, the maximum coating speed is plotted for several switching times and three interruption lengths.

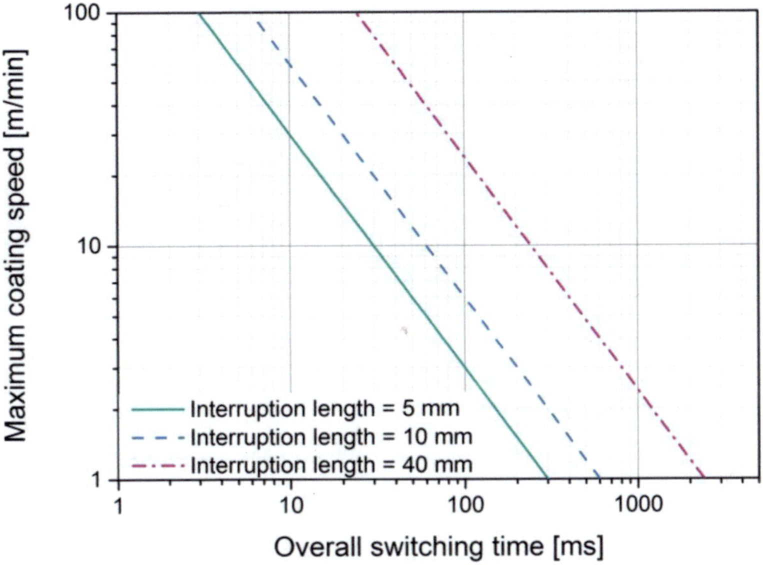

Figure 5.9: Maximum possible coating speed for several overall switching times at three interruption lengths.

As the above chart shows, an interruption length of $l_I = 40\,mm$ at a typical industrial web-speed of $u_W = 30\,m/min$ requires switching times of less than $80\,ms$.

5.2.2 Experimental results and discussion

To compare the theoretical and experimental analyses, the commanded switching state of the ball valve was measured at the same time as the system pressure. The results for $u_W = 5\,m/min,\ l_C = 60\,mm$ *and* $l_I = 40\,mm$ are plotted in Figure 5.10.

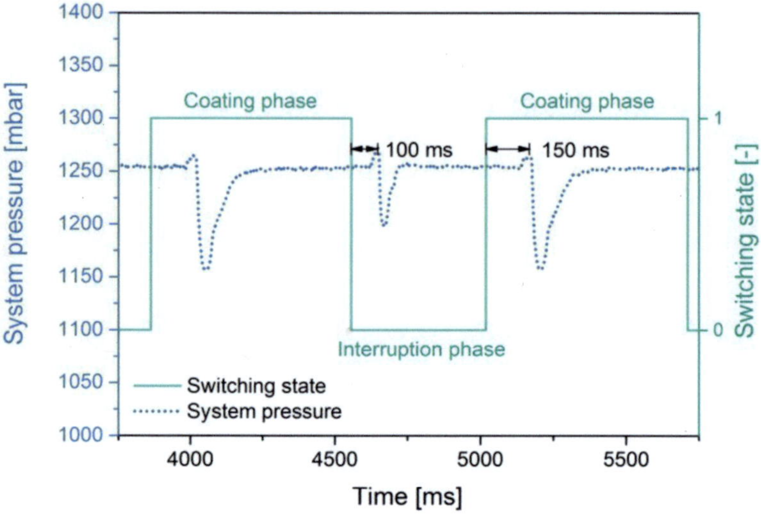

Figure 5.10: Commanded switching state of the bypass-ball-valve and resulting system pressure in LIB anode slurry 2. The coating was realized at $u_W = 5\ m/min$, $l_C = 60\ mm$, $l_I = 40\ mm$.

The above results show a delayed reaction of the liquid pressure to the commanded switching moment. At the end of the coating phase, there is a $100\,ms$ delay before the valve shut-off causes a slight pressure increase. After this short peak, the system pressure falls and then ramps back up as described in 5.1.3. Due to its design, the liquid flow is briefly hindered by the turning ball. This results in the small pressure peaks. At the spring-supported end of the interruption phase, there is a $150\,ms$ delay before the pressure peak is measured.

A short time delay may be attributed to the pressurizeed air valves ($10/12\,ms$ (Festo AG & Co. KG, 2012)) that effect the turning of the bypass-valve. Nevertheless, the $90°$ rotation of the ball valve probably consumes the majority of the time. The shortest interruption phase achievable was $t = 250\ ms$. The results in Figure 5.9 imply that the ball valve used here limits the coating speed to $u_W \leq 10\,m/min$ for an interruption length of $l_I = 40\,mm$. In contrast, the process windows presented in chapter 3 show no limitation up to $u_W \approx 100\,m/min$ for films with $h = 100\,\mu m$. If higher coating speeds are required, then faster bypass-valves—for example, piston-valves, with only a 1-dimensional movement—must be used instead.

In other intermitting concepts, mechanical movements of the die or the web achieve sharp edges and a clear interruption. In this case, as well, the above considerations about time-consuming movements are valid. Additionally, the increased impulse at higher coating speed and the accompanying system-induced vibrations should be kept in mind.

6 Conclusions

Large-scale, secondary lithium-ion batteries could be a key technology to compensate for the fluctuating energy supply of renewable sources. However, their manufacture is still very costly. In the processing of Lithium-ion batteries, the slot die coating of the electrodes is a crucial step, one that is not yet fully understood.

Therefore, the diverse requirements and challenges associated with this coating method were discussed in this work. The usual lithium-ion battery electrode slurries are highly particle loaded and highly viscous, with strong, shear-thinning behavior, which makes their use as coating materials a challenging endeavor.

Within the individual chapters "Process windows", "Edge formation" and "Intermittent slot die coating" models were developed and experimentally validated to help to optimize slot die coating processes.

6.1 Process windows

In this work, air entrainment was observed to be the predominating factor in determining the minimum film thickness limit for single-layer LIB coatings. This was so, even at higher coating speeds. To predict this limit, existing break-up models were discussed and applied. The model predictions for the LIB-electrode slurries were accompanied by those for a Newtonian liquid to reveal rheological singularities. The results agreed well with experimental observations and might be used to help increase current production speeds. It was shown that relatively thick films with gap ratios close to one are not limited by air entrainment at ambient conditions.

Subdivided LIB electrode might improve overall cell-performance. However, even for bilayer slot dies, it is reasonable that the altered gap regime also highly affects the processing limits. Thus, to predict the process limits for bilayer slot die coating, the above-mentioned model was refined by the author. To characterize the impact of the additional mid-lip pressure gradient, a top to bottom film height ratio M was introduced. This ratio is able to switch the signs of the additional pressure gradient beneath the mid-lip. As confirmed by the experimental results for both Newtonian and shear-thinning coating media, the additional pressure gradient increases the minimum film thickness at a top to bottom feed ratio of one or larger. However, by choosing a critical ratio M_{crit}, this effect can be eliminated.

6.2 Coating edge formation

During slot die coating, unfavorable super-elevations can occur at the film edges. In the downstream LIB-cell assembly, these so-called heavy edges cause undesirable folds and lead to inhomogeneous electrode stacks. To date, the literature only contains a model for super-elevations in hot melt film-extrusions. In this work, this model based on deformation stresses was adapted and extended for slot die coating. According to the theory, the liquid is accelerated between the gap and the web-speed plug flow. While the upper part of the film undergoes a neck-in, surface-adhesion pins the lower part, which causes super-elevated edges to be formed. Since, this adhesion to a substrate is not present in film-extrusion an adhesion-parameter λ was invented.

To validate the model, LIB electrode films were coated for different gap ratios and, thus, different accelerations. The results showed a strong increasing trend of edge heights towards higher gap ratios. An additional series for different coating speeds, as predicted, showed no dependence in the applied range.

The composition of the adhesion parameter λ is yet unknown, but has to be dependend to the given liquid viscosity and thus, the gap ratio. Therefore, the parameter was fitted for the applied liquid system and the adjusted gap ratios to the experimental results. With some minor deviations, the model describes the observed trend for the given gap ratios quite well. For gap ratios approaching unity, the calculations predict smaller edge super-elevations than observed. This can be related to the model-assumptions of an averaged gap velocity.

Thus, it was possible to validate this half-empirical model for the prediction of super-elevated edge heights in LIB electrode slot die coating. This model can help to lower the production wastage of coating lines, by decreasing heavy edges to an acceptable amount.

6.3 Intermittent slot die coating

Intermittent slot die coating is an important process step for manufacturing large-scale lithium-ion battery electrodes. Although there are many patents for intermittent film-forming equipment, the dynamic part of this coating science has not yet been covered by commonly-available scientific research.

In this work, an experimental set-up was established to investigate the dominating and limiting processes behind intermittent slot die coating on a very common basis. The flow of a LIB slurry was controlled through a precise syringe pump and interrupted by switching a 3-way bypass-valve between a waste pipe and a slot die.

Literature and experiments by the author showed a direct relation between liquid pressure in the coating feed system and the casted film. Hence, any film optimization requires a controlled die pressure. For casting a homogeneous film, this pressure has to reach a steady state with a compulsory leveling at each start. To shorten this ramping period, a non-Newtonian pressure calculation model for each steady state was developed to align the system pressures of the coating and interruption phases. This resulted in an approximately constant system pressure and the predicted film thickness. In contrast, a different system pressure during the coating and interruption phases resulted in too thin or too thick wet films. Thus, the system pressure upstream of the bypass-valve could not overcome the ramping within the coating length. The application of a Newtonian model system also showed this ramping behavior, although, here, it seems to be caused by omnipresent inertial forces. Hence intermittent slot die coating should always be accompanied by a prior pressure calculation.

In further considerations, the speed-limiting mechanisms of the intermittent coating methods and the details of the applied set-up were discussed. Experimental measurements revealed that the enforced interruption method limits the projected coating more than the relevant process window.

7 References

Barré, A., Deguilhem, B., Grolleau, S., Gérard, M., Suard, F., Riu, D., 2013. A review on lithium-ion battery ageing mechanisms and estimations for automotive applications. Journal of Power Sources 241, 680–689. 10.1016/j.jpowsour.2013.05.040.

Baumeister, M., Fleischer, J., 2014. Integrated cut and place module for high productive manufacturing of lithium-ion cells. CIRP Annals - Manufacturing Technology 63 (1), 5–8. 10.1016/j.cirp.2014.03.063.

Baunach, M., 2014. Influence of drying rate on Li-ion battery anodes, 7 September 2014, San Diego CA, USA.

Baunach, M., Jaiser, S., Schmelzle, S., Nirschl, H., Scharfer, P., Schabel, W., 2015. Delamination behavior of lithium-ion battery anodes: Influence of drying temperature during electrode processing. (accepted). Drying Technology.

Benhamou, E., 2011. Advanced Coating and Drying Techniques: Innovation and Technology - The Keys to Success. Advanced Automotive Battery Conference. Advanced Automotive Battery Conference, 6 June 2011, AABC 2011, Mainz, Germany.

Bhamidipati, K.L., Didari, S., Bedell, P., Harris, T.A.L., 2011. Wetting phenomena during processing of high-viscosity shear-thinning fluid. Journal of Non-Newtonian Fluid Mechanics 166 (12), 723–733.

Bitsch, B., Dittmann, J., Schmitt, M., Scharfer, P., Schabel, W., Willenbacher, N., 2014. A novel slurry concept for the fabrication of lithium-ion battery electrodes with beneficial properties. Journal of Power Sources 265, 81–90. 10.1016/j.jpowsour.2014.04.115.

Bitsch, B., Willenbacher, N., Wenzel, V., Schmelzle, S., Nirschl, H., 2015. Impact of Mechanical Process Engineering on the Fabrication Process of Electrodes for Lithium Ion Batteries: Einflüsse der mechanischen Verfahrenstechnik auf die Herstellung von Elektroden für Lithium-

Ionen-Batterien. Chemie Ingenieur Technik 87 (4), 466–474. 10.1002/cite.201400093.

Böhme, G., 2000. Strömungsmechanik nichtnewtonscher Fluide: Mit 49 Aufgaben, 2nd ed. Teubner, Stuttgart [u.a.].

Briec Emma, Müller Beate, 2015. Electric Vehicle Batteries: Moving from Research towards Innovation: Reports of the PPP European Green Vehicles Initiative, 2015th ed. Springer International Publishing, Cham.

Broom, K., 2011. Lithium-Ion Cell Manufacturing. Advanced Automotive Battery Conference. Advanced Automotive Battery Conference, 6 June 2011, AABC 2011, Mainz, Germany.

Bürkin, C., 2012. Untersuchung von Randeffekten bei der Beschichtung von Elektroden für Lithium-Ionen Batterien: Characterization of slot-die coated lithium-ion battery electrodes with special attention to edge effects. Bachelor Thesis, Karlsruhe, Germany.

Carvalho, M.S., Kheshgi, H.S., 2000. Low-flow limit in slot coating: Theory and experiments. AIChE Journal 46 (10), 1907–1917. 10.1002/aic.690461003.

Chang, Y.-R., Chang, H.-M., Lin, C.-F., Liu, T.-J., Wu, P.-Y., 2007. Three minimum wet thickness regions of slot die coating. Journal of Colloid and Interface Science 308 (1), 222–230. 10.1016/j.jcis.2006.11.054.

Chang, Y.-R., Lin, C.-F., Liu, T.-J., 2009. Start-up of slot die coating. Polymer Engineering & Science 49 (6), 1158–1167. 10.1002/pen.21360.

Chen, Y.-H., Wang, C.-W., Zhang, X., Sastry, A.M., 2010. Porous cathode optimization for lithium cells: Ionic and electronic conductivity, capacity, and selection of materials. Journal of Power Sources 195 (9), 2851–2862. 10.1016/j.jpowsour.2009.11.044.

Cholinski, E.J., 1993. Apparatus for patch coating printed circuit boards US 5183508 A.

Chu, W.-B., Yang, J.-W., Wang, Y.-C., Liu, T.-J., Tiu, C., Guo, J., 2006. The effect of inorganic particles on slot die coating of poly(vinyl alcohol) solutions. Journal of Colloid and Interface Science 297 (1), 215–225. 10.1016/j.jcis.2005.10.056.

Cohen, E.D., Gutoff, E.B., 1992. Modern Coating and Drying Technology. John Wiley & Sons.

Consultants, Roland Berger Strategy, 2011a. Powertrain 2020-The Li-Ion Battery Value Chain – Trends and implications.

Consultants, Roland Berger Strategy, 2011b. Zukunftsfeld Elektromobilität: Chancen und Herausforderungen für den deutschen Maschinen- und Anlagenbau.

Daniel, C., 2008. Materials and processing for lithium-ion batteries. JOM 60 (9), 43–48. 10.1007/s11837-008-0116-x.

Deutscher Bundestag, 2000. Gesetz für den Vorrang Erneuerbarer Energien (Erneuerbare-Energien-Gesetz – EEG) sowie zur Änderung des Energiewirtschaftsgesetzes und des Mineralölsteuergesetzes: EEG 2000.

Deutscher Bundestag, 2004. Gesetz zur Neuregelung des Rechts der Erneuerbaren Energien im Strombereich: EEG 2004.

Deutscher Bundestag, 2008. Gesetz zur Neuregelung des Rechts der Erneuerbaren Energien im Strombereich und zur Änderung damit zusammenhängender Vorschriften: EEG 2009.

Deutscher Bundestag, 2011. Gesetz zur Neuregelung des Rechtsrahmens für die Förderung der Stromerzeugung aus erneuerbaren Energien: EEG 2012.

Deutscher Bundestag, 2012. Gesetz zur Änderung des Rechtsrahmens für Strom aus solarer Strahlungsenergie und zu weiteren Änderungen im Recht der erneuerbaren Energien: EEG 2012 II.

Deutscher Bundestag, 2014. Gesetz zur grundlegenden Reform des Erneuerbare-Energien-Gesetzes und zur Änderung weiterer Bestimmungen des Energiewirtschaftsrechts: EEG 2014.

Diehm, R., 2014. Untersuchung der Skalierbarkeit von mehrlagigen und intermittierten Elektrodenbe-schichtungen für Lithium-Ionen-Batterien: Investigations on scalability of intermittent and multilayer coated lithium-ion battery electrodes. Diploma Thesis, Karlsruhe, Germany.

Dobroth, T., Erwin, L., 1986. Causes of edge beads in cast films. Polymer Engineering & Science 26 (7), 462–467. 10.1002/pen.760260704.

Dominko, R., Gaberscek, M., Drofenik, J., Bele, M., Pejovnik, S., Jamnik, J., 2003a. The role of carbon black distribution in cathodes for Li ion batteries. Journal of Power Sources 119, 770–773.

Dominko, R., Gaberšček, M., Drofenik, J., Bele, M., Jamnik, J., 2003b. Influence of carbon black distribution on performance of oxide cathodes for Li ion batteries. Electrochimica Acta 48 (24), 3709–3716.

Durst, F., Dongari, N., Sambasivam, R., 2007. Slot coaters operating in their bead mode. Coating 11/2007, 1–6.

Durst, F., Wagner, H.-G., 1997. Slot coating, in: Kistler, S.F., Schweizer, P.M. (Eds.), Liquid film coating: scientific principles and their technological implications. Chapman & Hall, pp. 401–426.

Ebert, L., Roscher, M.A., Wolter, M., 2014. Schnellere Serienproduktion von Lithium-Ionen-Batterien. ATZ Automobiltech Z 116 (4), 60–63. 10.1007/s35148-014-0383-5.

Eichinger, G., Semrau, G., 1990. Lithiumbatterien I. Chemische Grundlagen. Chem. Unserer Zeit 24 (1), 32–36. 10.1002/ciuz.19900240108.

Festo AG & Co. KG (Ed.), 2012. Product documentation: VOVG magnetic valves.

Fleischer, J., Ruprecht, E., Baumeister, M., Haag, S., 2012. Automated Handling of Limp Foils in Lithium-Ion-Cell Manufacturing, in: Dornfeld, D.A., Linke, B.S. (Eds.), Leveraging Technology for a Sustainable World. Springer Berlin Heidelberg, Berlin, Heidelberg, pp. 353–356.

Fraunhofer-Institut für System- und Innovationsforschung ISI, 2013. Energiespeicher Monitoring für die Elektromobilität (EMOTOR): Bericht zur Produktion und Ökobilanzierung.

Gates, I.D., 1999. Slot coating flows: feasibility, quality, Minneapolis.

Gerischer, H., Decker, F., Scrosati, B., 1994. The electronic and the ionic contribution to the free energy of alkali metals in intercalation compounds. J. Electrochem. Soc. 141 (9), 2297–2300.

German Federal Government, 2009. German Federal Government's National Electromobility Development Plan.

German Federal Government, 2010. The Federal Government's energy concept of 2010.

Glawe, A., 2014. From theory to practice with pilot coating trials. Karlsruhe Institute of Technology, Institute of Thermal Process Engineering -Thin Film Technology. Short Course - Coating and Drying of Thin Films, 19 May 2014, Karlsruhe, Germany.

Gulbinska, M.K., 2014. Lithium-ion Battery Materials and Engineering: Current Topics and Problems from the Manufacturing Perspective, 2014th ed. Springer London, London.

Gutoff, E.B., Cohen, E.D., Kheboian, G.I., 2006. Coating and drying defects: troubleshooting operating problems. Wiley-Interscience.

Habermann, R., 2011. Einfluss der Beanspruchungsmechanismen und -intensitäten auf die Produkteigenschaften beim Feststoffmischen. Chemie Ingenieur Technik 83 (5), 689–698. 10.1002/cite.201000154.

Hagiwara, H., 2012. Study of the Binder Distribution in an Electrode During Dryin, 9 September 2012, Atlanta GA, USA.

Haselrieder, W., Ivanov, S., Christen, D.K., Bockholt, H., Kwade, A., 2013. Impact of the Calendering Process on the Interfacial Structure and the Related Electrochemical Performance of Secondary Lithium-Ion Batteries. ECS Transactions 50 (26), 59–70. 10.1149/05026.0059ecst.

Hau, E., 2008. Wind turbines: fundamentals, technology, applications, cost-efficiency. Springer, Berlin.

Higgins, B.G., Scriven, L.E., 1980. Capillary pressure and viscous pressure drop set bounds on coating bead operability. Chemical Engineering Science 35 (3), 673–682. 10.1016/0009-2509(80)80018-2.

Horiuchi, R., Suszynski, W.J., Carvalho, M.S., 2015. Simultaneous multilayer coating of water-based and alcohol-based solutions. J Coat Technol Res, 1–8. 10.1007/s11998-015-9689-9.

Hrymak, A.N., Taylor, D., 1999. Visualization and flow simulation of a two-layer slot coater. Chemical Engineering Science 54, 909–918.

Huggins, R.A., 2009. Advanced batteries: Materials science aspects. Springer, New York, [London].

Huth, A., Sekler, R., 2014. Switchable slot valve for a coating system, coating system and coating method DE 102012217683 B3.

Imachi, N., Takano, Y., Fujimoto, H., Kida, Y., Fujitani, S., 2007. Layered Cathode for Improving Safety of Li-Ion Batteries. J. Electrochem. Soc. 154 (5), A412. 10.1149/1.2709503.

International Energy Agency, 2014. Energy Storage: Technology Roadmap. OECD Publishing, Paris.

Iwashita, Y., Endo, S., Morimoto, K., 1999. Intermittent coating method and apparatus therefor US 5989622 A.

Jaiser, S., 2015. Identification of morphology-critical stages during drying of particulate electrode slurries, 28 April 2015, Aachen, Germany.

Jang, I., Song, S., 2013. A model for prediction of minimum coating thickness in high speed slot coating. International Journal of Heat and Fluid Flow 40, 180–185. 10.1016/j.ijheatfluidflow.2013.01.002.

Janssen, F., 2011. Coating nozzle for intermittent application of fluid medium to substrate, has sealing unit arranged in channel, so that channel is opened and closed to produce intermittent application of medium in transporting direction DE 102010017965 A1.

Junges, P., 2011. Untersuchung des Beschichtungsverhaltens von Elektrodenpasten für Lithium-Ionen-Batterien: Investigation of the coating behavior of lithium-ion battery electrode slurries. Diploma Thesis, Karlsruhe, Germany.

Kaiser, J., Wenzel, V., Nirschl, H., Bitsch, B., Willenbacher, N., Baunach, M., Schmitt, M., Jaiser, S., Scharfer, P., Schabel, W., 2014. Prozess- und Produktentwicklung von Elektroden für Li-Ionen-Zellen Process and Product Development of Electrodes for Lithium-Ion Cells. Chemie Ingenieur Technik, n/a-n/a. 10.1002/cite.201300085.

Kamişli, F., Ryan, M.E., 1999. Perturbation method in gas-assisted power-law fluid displacement in a circular tube and a rectangular channel. Chemical Engineering Journal 75 (3), 167–176. 10.1016/s1385-8947(99)00088-1.

Ketterer B., U. Karl, D. Möst, S. Ulrich, 2009. Lithium-Ionen Batterien: Lithium-Ion Batteries: State of the Art and Application Potential in

Hybrid-, Plug-In Hybrid- and Electric Vehicles. Forschungszentrum Karlsruhe in der Helmholtz-Gemeinschaft/ Wissenschaftliche Berichte FZKA 7503.

Kim, K.M., Jeon, W.S., Chung, I.J., Chang, S.H., 1999. Effect of mixing sequences on the electrode characteristics of lithium-ion rechargeable batteries. Journal of Power Sources 83 (1), 108–113.

Kim S, Kim JH, Ahn KH, Lee SJ, 2009. Rheological perspectives of industrial coating process. Korea-Australia Rheology Journal 21 (2), 83–89.

Kistler, S.F., Schweizer, P.M. (Eds.), 1997. Liquid film coating: scientific principles and their technological implications. Chapman & Hall.

KIT, 2014. World's Fastest Manufacture of Battery Electrodes, http://www.kit.edu/kit/english/pi_2014_15826.php.

Kitz, P., 2013. Untersuchung eines neuartigen Schlitzgießerkonzepts zur intermittierenden Beschichtung von Elektroden für Lithium-Ionen-Batterien: Charakterization of a new slot die design for intermittent lithium-ion battery coatings. Bachelor Thesis, Karlsruhe, Germany.

Koh, H., Kwon, I., Jung, H., Hyun, J., 2012. Operability window of slot coating using viscocapillary model for Carreau-type coating liquids. Korea-Australia Rheology Journal 24 (2), 137–141. 10.1007/s13367-012-0016-z.

Konrad, M., 2013. Photovoltaik. Carl Hanser Verlag GmbH & Co. KG.

KRÜSS GmbH. Pendant drop. URL: <http://www.kruss.de/services/education-theory/glossary/pendant-drop/>. Accessed 18 June 2015.

Kurfer, J., Westermeier, M., Tammer, C., Reinhart, G., 2012. Production of large-area lithium-ion cells – Preconditioning, cell stacking and quality assurance. CIRP Annals - Manufacturing Technology 61 (1), 1–4. 10.1016/j.cirp.2012.03.101.

Landau, L., Levich B., 1942. Dragging of a Liquid by a Moving Plate. Acta Physicochimica U.R.S.S. VOL. XVII (1-2), 42–54.

Lee, K.-Y., Liu, L.-D., Ta-Jo, L., 1992. Minimum wet thickness in extrusion slot coating. Chemical Engineering Science 47 (7), 1703–1713.

Lee, S.H., Koh, H.J., Ryu, B.K., Kim, S.J., Jung, H.W., Hyun, J.C., 2011. Operability coating windows and frequency response in slot coating flows from a viscocapillary model. Chemical Engineering Science 66 (21), 4953–4959. 10.1016/j.ces.2011.04.044.

Lin, C.-F., Hill Wong, David S., Liu, T.-J., Wu, P.-Y., 2010. Operating windows of slot die coating: Comparison of theoretical predictions with experimental observations. Advances in Polymer Technology 29 (1), 31–44. 10.1002/adv.20173.

Linden, D., Reddy, T.B., 2002. Handbook of batteries, 3rd ed. McGraw-Hill, New York.

Liu, C., Li, F., Ma, L.-P., Cheng, H.-M., 2010. Advanced materials for energy storage. Advanced materials (Deerfield Beach, Fla.) 22 (8), E28-62. 10.1002/adma.200903328.

Lu, S.-Y., Lin, Y.-P., Liu, T.-J., 2001. Coating Window for Double Layer Extrusion Slot Coating of Poly(Vinyl-Alcohol) Solutions. Polymer Engineering and Science (Vol. 41, No. 10), 1823–1829.

Masuda, N., Watanabe, M., 2006. Intermittent coating apparatus and intermittent coating method US 7105203 B1.

McKinsey & Company, 2012. Battery technology charges ahead. McKinsey Quarterly 3, 5–50.

Mecklenburg, A., 2014. Industrial manufacturing of lithium-ion battery cells. Karlsruhe Institute of Technology, Institute of Thermal Process Engineering -Thin Film Technology. Short Course - Coating and Drying of Thin Films, 19 May 2014, Karlsruhe, Germany.

Menges, G., 2002. Werkstoffkunde Kunststoffe, 5th ed. Hanser, München, Wien.

Mezger, T., 2010. Das Rheologie-Handbuch: Für Anwender von Rotations- und Oszillations-Rheometern, 3rd ed. Vincentz Network, Hannover.

Moskon, J., Dominko, R., Cerc-Korosec, R., Gaberscek, M., Jamnik, J., 2007. Morphology and electrical properties of conductive carbon coatings for cathode materials. Journal of Power Sources 174 (2), 683–688. 10.1016/j.jpowsour.2007.06.239.

Nam, J., Carvalho, M.S., 2009. Mid-gap invasion in two-layer slot coating. Journal of Fluid Mechanics 631, 397. 10.1017/s0022112009007022.

Nam, J., Carvalho, M.S., 2010. Linear stability analysis of two-layer rectilinear flow in slot coating. AIChE Journal 56 (10), 2503–2512. 10.1002/aic.12172.

Nazri, G., Pistoia, G., 2009. Lithium batteries: Science and technology. Springer, New York, N.Y.

Neumann, G., 2011. Lithium Ion Secondary Batteries: State of the Art and Future Developments. Lithium-Akkumulatoren - Stand der Technik und zukünftige Entwicklungen. Chemie Ingenieur Technik 83 (11), 2042–2050. 10.1002/cite.201100075.

Ozawa, K., 2009. Lithium ion rechargeable batteries: Materials, technology, and applications. (E-Book). Wiley-VCH, Weinheim.

Pistoia, G., 2014. Lithium-ion batteries: Advances and applications, 1st ed. Elsevier, Amsterdam, The Netherlands.

Rolinger, L., 2013. Untersuchung von Randeffekten bei der Beschichtung von Anoden und Kathoden für Lithi-um-Ionen-Batterien: Characterization of slot-die coated lithium-ion battery anodes and cathodes with special attention to edge effects. Bachelor Thesis, Karlsruhe, Germany.

Romero, O., 2004. Low-flow limit in slot coating of dilute solutions of high molecular weight polymer. Journal of Non-Newtonian Fluid Mechanics 118 (2-3), 137–156. 10.1016/j.jnnfm.2004.03.004.

Romero, O.J., Scriven, L.E., Carvalho, M.S., 2006. Slot coating of mildly viscoelastic liquids. Journal of Non-Newtonian Fluid Mechanics 138 (2-3), 63–75. 10.1016/j.jnnfm.2005.11.010.

Romero, O.J., Suszynski, W.J., Scriven, L.E., Carvalho, M.S., 2004. Low-flow limit in slot coating of dilute solutions of high molecular weight polymer. Journal of Non-Newtonian Fluid Mechanics 118 (2-3), 137–156. 10.1016/j.jnnfm.2004.03.004.

Ruprecht, E., Kaiser, J., 2013. Kostenoptimierte Batterie-Zellfertigung. ATZ Elektron 8 (3), 180–185. 10.1365/s35658-013-0292-5.

Ruschak, K.J., 1976. Limiting flow in a pre-metered coating device. Chemical Engineering Science 31 (11), 1057–1060.

Russner, N., 2012. Simulation der Fluidströmung innerhalb von Schlitzgießern während der Beschichtung von Elektroden für Lithium-Ionen-Batterien: Simulation of the interior liquid flow of slot dies during the coating of lithium-ion battery electrodes. Bachelor Thesis, Karlsruhe, Germany.

Sakai, Y., Yoshikawa, T., Ide, A., 2002. Intermittent coating system and intermittent coating method US 6455105 B1.

Sartor, L. Slot coating: Fluid mechanics and die design, Minneapolis.

Schlögl, R., 2013. Chemical energy storage. De Gruyter, Berlin.

Schlünder, E.-U., 1996. Einführung in die Stoffübertragung: Mit 6 Tabellen. Vieweg, Braunschweig, Wiesbaden.

Schmitkons, J.W., Turner, J., Zupan, M.P., Rivas, A., Benecke, J., Cieplik, A., Burmester, T., Boger, B., 1998. Snuff back controlled coating dispensing apparatus and methods US 5733597 A.

Schmitt, M., Baunach, M., Scharfer, P., Schabel, W., 2012a. Wet film characterization of slot die coated multilayer electrodes: (Talk). 12-9-2012. Karlsruhe Institute of Technology. 16th International Coating Science and Technology Symposium, 2012, Atlanta, GA, USA.

Schmitt, M., Baunach, M., Scharfer, P., Schabel, W., 2012b. Surface quality of slot die coated battery electrodes with special attention to edge effects: (Poster). Karlsruhe Institute of Technology. 16th International Coating Science and Technology Symposium, 12 September 2012, Atlanta, GA, USA.

Schmitt, M., Baunach, M., Wengeler, L., Peters, K., Junges, P., Scharfer, P., Schabel, W., 2013a. Slot-die processing of lithium-ion battery electrodes—Coating window characterization. Chemical Engineering and Processing: Process Intensification 68 (0), 32–37. 10.1016/j.cep.2012.10.011.

Schmitt, M., Carvalho, M.S., Scharfer, P., Schabel, W., 2013b. Break-up phenomena in multilayer slot die coating of lithium-ion batteries:

(Poster). Karlsruhe Institute of Technology. European Coating Symposium 2013, 2013, Mons, Belgium.

Schmitt, M., Diehm, R., Scharfer, P., Schabel, W., 2015a. An experimental and analytical study on intermittent slot die coating of viscoelastic battery slurries. (Accepted). J Coat Technol Res. 10.1007/s11998-015-9717-9.

Schmitt, M., Kitz, P., Scharfer, P., Schabel, W. A half-empirical model to predict heavy edges in slot die coating. (to be submitted).

Schmitt, M., Raupp, S., Wagner, D., Scharfer, P., Schabel, W., 2015b. Analytical determination of process windows for bilayer slot die coating. (Accepted). J Coat Technol Res. 10.1007/s11998-015-9701-4.

Schmitt, M., Scharfer, P., Schabel, W., 2013c. Slot die coating of lithium-ion battery electrodes: investigations on edge effect issues for stripe and pattern coatings. J Coat Technol Res, 1–7. 10.1007/s11998-013-9498-y.

Schmitt, M., Scharfer, P., Schabel, W., 2013d. Stripe and pattern coatings of slot die coated battery electrodes – investigation of edge effects (Talk). Karlsruhe Institute of Technology. European Coating Symposium 2013, 8 November 2013, Mons, Belgium.

Schneider, R., 2012. Untersuchung des simultanen Mehrlagenauftrags von Elektrodenpasten für Lithium-Ionen-Batterien: Investigation on multilayer coating of lithium-ion battery electrode slurries. Student Research Thesis, Karlsruhe, Germany.

Schuster, M., 2015. Untersuchung und Hochgeschwindigkeitsautomatisierung intermittierter Schlitzdüsenbe-schichtungen: Investigations and high speed automation of intermittent slot die coating. Diploma Thesis, Karlsruhe, Germany.

Secor, R.B., 1997. Analysis and design of internal coating die cavities, in: Kistler, S.F., Schweizer, P.M. (Eds.), Liquid film coating: scientific principles and their technological implications. Chapman & Hall, pp. 369–398.

Shinozaki, K., Sakai, Y., 2003. Coating apparatus for applying coating material intermittently to substrate.

Spiegel, S., 2014. Untersuchung des Systemdrucks bei intermittierten Elektrodenbeschichtungen für Lithium-Ionen-Batterien: Investigations on the system pressure at intermittent lithium-ion battery electrode coatings. Bachelor Thesis, Karlsruhe, Germany.

Szczepaniak, W., Janssen, F., Blum, J., 2014. Vorrichtung zum präzisen linearen Verstellen eines Maschinenteils DE 102012110305 A1.

Taylor, D., 1997. Two-Layer Slot Coating: Study of Die Geometry and Interfacial Region, Hamilton, Ontario.

Terashita, K., Miyanami, K., 2002. Kneading and dispersion of positive electrode materials in a lithium ion secondary battery for high-density film. Advanced Powder Technology 13 (2), 201–214. 10.1163/156855202760166541.

Thackeray, M.M., Wolverton, C., Isaacs, E.D., 2012. Electrical energy storage for transportation - approaching the limits of, and going beyond, lithium-ion batteries. Energy & Environmental Science. 10.1039/c2ee21892e.

Tran, H.Y., 2013. Verfahrenstechnische Entwicklung und Untersuchung von Elektroden und deren Herstellprozess für innovative Lithium-Hochleistungsbatterien.

Tsuda, T., 2010. Coating Flows of Power-Law Non-Newtonian Fluids in Slot Coating. Nihon Reoroji Gakk 38 (4-5), 223–230.

Unger, S., 2013. Untersuchung von Prozesseinflüssen bei der Herstellung von Elektroden für Lithium-Ionen-Batterien durch Schlitzdüsen: Investigations on process conditions for slot die casted lithium-ion battery electrodes. Diploma Thesis, Karlsruhe, Germany.

Van Schalkwijk, Walter A, Scrosati, B., op. 2002. Advances in lithium-ion batteries. Kluwer Academic/Plenum Publishers, New York, Boston, Dordrecht [etc.].

VDI-Gesellschaft Verfahrenstechnik und Chemieingenieurwesen, Gesellschaft, V.D., 2010. VDI Heat Atlas. Springer.

Vetter, J., Novák, P., Wagner, M.R., Veit, C., Möller, K.-C., Besenhard, J.O., Winter, M., Wohlfahrt-Mehrens, M., Vogler, C., Hammouche, A.,

2005. Ageing mechanisms in lithium-ion batteries. Journal of Power Sources 147 (1-2), 269–281. 10.1016/j.jpowsour.2005.01.006.

Vogt, M., 2012. Untersuchung eines neuartigen Schlitzgießerkonzepts zur intermittierenden Beschichtung von Elektroden für Lithium-Ionen-Batterien: Investigation on a newly developed slot die concept for intermittent lithium-ion battery electrode coatings. Bachelor Thesis, Karlsruhe, Germany.

Vries, I. de, 2014. Intermittent slot die coating of low viscous soulutions. (Talk). Holst Center, Eindhoven, NL. 17th International Coating Science and Technology Symposium (ISCST 2014), 9 September 2014, San Diego, CA, USA.

Wagner, D., 2012. Untersuchung von Abrisskriterien bei der simultanen Mehrlagenbeschichtung von Elektroden für Lithium-Ionen Batterien: Investigation on coating limits of multilayer coated lithium-ion battery electrodes. Bachelor Thesis, Karlsruhe, Germany.

Watanabe, M., Ueyama, Y., Nakamura, T., Ohana, Y., Hayashi, T., 1996. Intermittent coating apparatus, intermittent coating method and manufacturing method of battery electrodes, and non aqueous electrolyte cell EP 0720872 B1.

Weinstein, S.J., Ruschak, K.J., 2004. Coating Flows. Annual Review of Fluid Mechanics 36 (1), 29–53. 10.1146/annurev.fluid.36.050802.122049.

Weiß, M., 2013. Simulation der Strömung im Beschichtungsspalt von Mehrlagenschlitzdüsen während der Beschichtung von Elektroden für Lithium-Ionen Batterien: Simulation of liquid flow in the coating gap during multilayer slot-die coating of lithium-ion battery electrodes. Bachelor Thesis, Karlsruhe, Germany.

Wengeler, L., 2014. Coating and drying processes for functional films in polymer solar cells-from laboratory to pilot scale. KIT Scientific Publishing.

Wengeler, L., Schmitt, M., Peters, K., Scharfer, P., Schabel, W., 2013. Comparison of large scale coating techniques for organic and hybrid films in polymer based solar cells. Chemical Engineering and Processing: Process Intensification 68, 38–44.

Wenzel, V., Moeller, R.S., Nirschl, H., 2014. Influence of Mixing Technology and the Potential to Modify the Morphological Properties of Materials used in the Manufacture of Lithium-Ion Batteries. Energy Technology 2 (2), 176–182. 10.1002/ente.201300091.

Wittek, M., 2012. Simulation von Strömungsabrissen beim Beschichten von Lithium-Ionen Batterien: Simulation of liquid flow in the coating gap during slot-die coating of lithium-ion battery electrodes. Bachelor Thesis, Karlsruhe, Germany.

Yang, C.K., Wong, D.S., Liu, T.J., 2004a. An experimental study on discrete patch coating. Ind. Coat. Res 5, 43–58.

Yang, C.K., Wong, D. S. H., Liu, T.J., 2004b. The effects of polymer additives on the operating windows of slot coating. Polymer Engineering & Science 44 (10), 1970–1976. 10.1002/pen.20200.

Yoshio, M., Brodd, R.J., Kozawa, A., 2009. Lithium-ion batteries: science and technologies. Springer.

Youn, S.I., Kim, S.Y., Shin, D.M., Lee, J.S., Jung, H.W., Hyun, J.C., 2006. A review on viscocapillary models of pre-metered coating flows. Korea-Australia Rheology Journal 18 (4), 209–215.

Yu, Y.L., 1995. Reduction of the minimum wet thickness in extrusion slot coating.

Yuan, X., Liu, H., Zhang, J., 2012. Lithium-ion batteries: Advanced materials and technologies. CRC Press, Boca Raton, FL.

Zheng, H., Tan, L., Liu, G., Song, X., Battaglia, V.S., 2012. Calendering effects on the physical and electrochemical properties of Li[Ni1/3Mn1/3Co1/3]O2 cathode. Journal of Power Sources 208 (0), 52–57. 10.1016/j.jpowsour.2012.02.001.

8 Appendix

8.1 Permisson for the publication of thesis extracts

„Von: Hubbuch, Juergen (BLT)
Gesendet: Montag, 4. Mai 2015 17:48
An: Horn, Harald (EBI); Kreuter, Veronika (CIW)
Betreff: FW: Erlaubnis Auszugsveröffentlichungen

OK mit mir :-)

Mit TouchDown von meinem Android-Telefon gesendet (www.nitrodesk.com)

-----Original Message-----
From: Schmitt, Marcel (TVT) [marcel.schmitt@kit.edu]
Received: Montag, 04 Mai 2015, 17:08
To: Hubbuch, Juergen (BLT) [juergen.hubbuch@kit.edu]
CC: Kreuter, Veronika (CIW) [veronika.kreuter@kit.edu]
Subject: Erlaubnis Auszugsveröffentlichungen
Sehr geehrter Herr Prof. Hubbuch,

unter meinem Doktorvater Herrn Prof. Schabel bin ich in den letzten Zügen meiner Dissertation und würde gerne noch drei Veröffentlichungen meiner Forschungsergebnisse zum Thema einreichen.

Es handelt sich dabei um die Titel:
"An experimental and analytical study on intermittent slot die coating of viscoelastic battery slurries" – in: Journal of Coating Technology

"Analytical determination of process windows for bilayer slot die coating" – in: Journal of Coating Technology
"A half-empirical model to predict heavy edges in slot die coating" – Voraussichtlich in: Chemical Engineering Science

Meine Dissertation möchte ich auf den in diesen Schriften thematisierten Ergebnissen aufzubauen.

In Ihrer Funktion als Dekan würde ich Sie deshalb höflichst um eine schriftliche Erlaubnis zu diesen „Auszugsveröffentlichungen" bitten.

Freundliche Grüße und vielen Dank vorab
Marcel Schmitt"

8.2 Lithium-ion batteries

Batteries are so-called galvanic cells, comprised of two electrodes, an electrolyte, and a separator. The electrodes consist of an electrochemically active material placed on a current collector that ensures an electrical connection. The separator divides the cell into two physically separated and electrically isolated half-cells, but permits charge balancing through high ion-permeability. The conversion of chemical energy into usable electrical energy is based on redox reactions in each half-cell. Here, ion oxidation takes place at the anode and reduction at the cathode. To enable ion diffusion within the whole cell, the electrolyte fills the free space between the electrodes (Nazri and Pistoia, 2009; Ozawa, 2009; Yoshio et al., 2009; Yuan et al., 2012).

Conceptually, there are three types of batteries: Primary batteries allow only one discharging process and are thus one-time usable. Secondary batteries, so-called accumulators, are based on chemically reversible reactions and are rechargeable through a current reversal. Tertiary batteries need an external fuel supply for continuous operation and are referred to as fuel cells. Thus, the lithium-ion batteries discussed in this work fall into the category of accumulators (Linden and Reddy, 2002; Schlögl, 2013). In the following section, the relevant performance parameters and state of the art of LIBs and their components will be discussed.

8.2.1 Functionality and layout of lithium-ion batteries

As active LIB materials, lithium-metal-oxides represent the positive and often graphite the negative electrode. During the discharging process, the Li-ions are disintercalated from the graphite crystal lattice and thus release electrons. The ions must then diffuse through the electrolyte to the cathode to be intercalated again, a process in which each ion consumes a free electron. The nascent electrical energy can be used by an external consumer. By applying an electrical source to the system, these mechanisms can be reversed, so as to charge the lithium-ion cell again

In equations (8.1) to (8.3) the half-cell reactions and the resulting reaction of the whole cell are described (Schlögl, 2013).

$$Li_xC_6 \xleftrightarrow{dis-/charging} xLi^+ + 6C + xe^- \tag{8.1}$$

$$Li_{1-x}MO_2 + xLi^+ + xe^- \xleftrightarrow{dis-/charging} LiMO_2 \tag{8.2}$$

$$Li_{1-x}MO_2 + Li_xC_6 \xleftrightarrow{dis-/charging} LiMO_2 + 6C \tag{8.3}$$

The electrodes of lithium-ion battery cells are based on particulate materials coated on a metallic substrate. Polymeric binder systems provide for adhesion and cohesion, and carbon black establishes the conductivity between active material particles and substrate. Graphite is a frequently applied active material for LIB anodes, where copper foils are used as current-arresting substrates (Yuan et al., 2012). Because the various lithium-metal-oxides used as active materials on the cathode side would react with copper, non-reactive aluminium substrates are applied as current collectors for the cathode (Yuan et al., 2012).

In commercially available LIB cells, the current-arresting substrates are coated with active material on both sides. This permits a layering of multiple galvanic elements within a single housing.

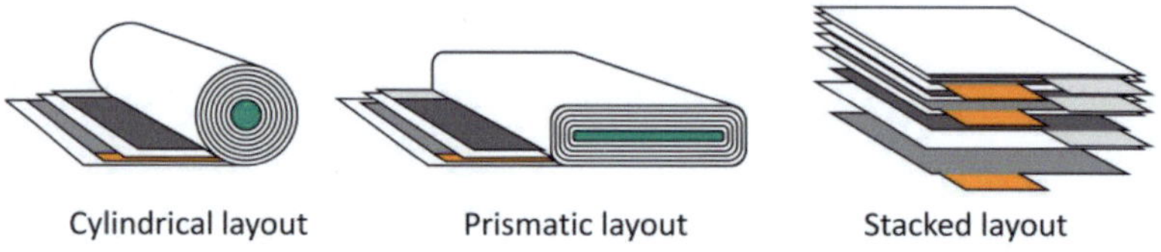

Figure 8.1: Commercially available lithium-ion battery cells are usually produced as cylindrical, prismatic, or stacked layouts.

The most common cell layouts are wound cylindrical, wound prismatic, and stacked (Linden and Reddy, 2002; Yoshio et al., 2009) (see Figure 8.1). Because cylindrical cells are more mechanically stable, they can withstand a higher inner pressure than flat layouts. On the other hand, flat layouts are easier to stack and offer better thermal control, albeit at higher costs (Van Schalkwijk, Walter A and Scrosati, op. 2002). Moreover, a stacked layout requires no mandrel and is thus very volume effective. The latest research in cell assembly combines all the advantages in a novel winding technology for stacked cells (KIT, 2014). Therefore, the electrodes manufactured in this work were designed for a stacked cell layout.

8.2.2 Performance parameters

There are four performance parameters that determine the efficiency of LIB layouts and their material combinations: cell voltage, specific storage capacity, energy density, and power density (Eichinger and Semrau, 1990; Ketterer B., U. Karl, D. Möst, S. Ulrich, 2009; Gulbinska, 2014).

The available cell voltage $\Phi^{Voltage}$, or electrode potential, results from the different potentials of the selected electrode materials and significantly affects cell performance. With no applied current, the cell voltage is proportional to the free reaction enthalpy and thus, to the difference in the applied chemical potentials. With an applied current, there are ohmic resistances, which lower these theoretical values (Gerischer et al., 1994). To compare various electrode materials, their potentials must be measured against those of a metallic lithium electrode, which is defined to be zero. The cell voltage of a specific combination of materials is then the difference in potential against metallic lithium. The common combination of $LiCoO_2$ as cathode and graphite as anode results in a cell voltage of $\Phi^{Voltage} \approx 3.6\,V$.

The specific storage capacity $\Phi^{Capacity}$ may be regarded as the amount of Li-ions available for the reversible intercalation or disintercalation process. There is a gravimetric specific storage capacity $\Phi^{Capacity}_{Grav}$ and a volumetric specific capacity $\Phi^{Capacity}_{Vol}$.

The energy density Φ^{Energy}_{Vol} or specific energy Φ^{Energy}_{Grav} is the product of the specific storage capacity and the cell voltage:

$$\Phi^{Energy}_{Vol/Grav} = \Phi^{Capacity}_{Vol/Grav} \cdot \Phi^{\mathrm{Voltage}} \qquad (8.4)$$

Hence, only a combination of high cell voltage and storage density allows for a preferably high energy density. In comparison to NiCd batteries for example, LIBs have a more than two times higher energy density and a more than three times higher specific energy (as shown in Figure 8.2). Thus, LIBs allow for smaller and lighter electricity storage, which makes them the prevailing technology in the handheld applications market.

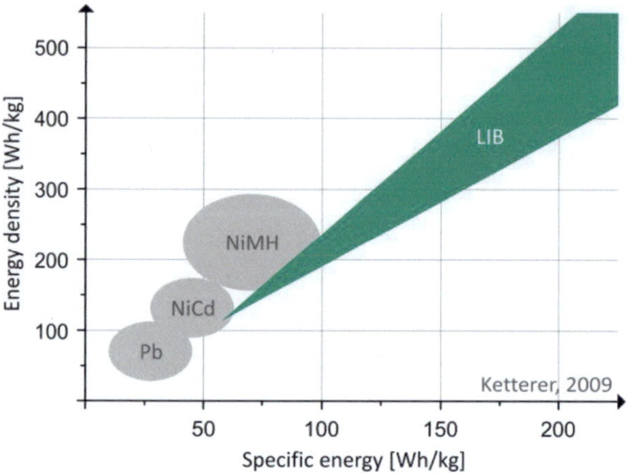

Figure 8.2: Schematic comparison of the specific energy and energy density of available battery technologies (Neumann, 2011).

Current-carrying capacity is crucial for high power density Φ_{Vol}^{Power} or specific power Φ_{Grav}^{Power}. The speed of the intercalation process is limited by electron and ion diffusion inside the electrode and electrolyte. By accepting low cell capacities, high power densities can be achieved by means of short diffusion paths in flat and small electrodes (Gulbinska, 2014). The so-called C-rate is used to define the charging or discharging current independent of the cell capacity (Linden and Reddy, 2002). The applied currents are defined as a fraction or a multiple part of the overall cell capacity $\Phi^{Capacity}$. Thus, a discharging rate of $1C$ for a cell with $\Phi^{Capacity} = 1000\ mAh$ results in a discharging current of $1000\ mA$ for one hour. Similar $2C$ results in a discharging rate of $2000\ mA$ for half an hour, and so on (Ketterer B., U. Karl, D. Möst, S. Ulrich, 2009).

8.2.3 State of the art electrode materials

Today's lithium-ion batteries are the prevailing storage technology in mobile handheld consumer applications, such as smart phones, laptops or tools. Within the past two decades, other technologies, such as nickel metal hydride batteries, have almost completely vanished. Nevertheless, new applications for LIBs in electric vehicles (EV) or in stationary energy storage devices for example, come along with new requirements. These requirements relate not only to lower weights, increased safety and recycleablility, but also to decreased material and manufacturing costs. In the long term, electrode materials should also exhibit greater cycle and calendaric durability at higher depths of discharging, high energy density, and increased temperature stability. All these properties depend on the selected electrode materials (Neumann, 2011).

The cathode materials of LIB are transition metal oxides and thus, functional ceramics. They possess characteristic crystal structures that provide the lithium-ion diffusion paths. The three different crystal structure types are one, two and three-dimensional (Nazri and Pistoia, 2009; Yoshio et al., 2009; Gulbinska, 2014; Pistoia, 2014). The one-dimensional, so-called olivine structure, is reported for $LiFePO_4$ (LFP) for example. Due to its limited linear structure, this material provides poor ion mobility but is extremely stable. In current applications, the restriced ion diffusion paths were shortened by the reduction of the particle size to nanoscales. To date, the most common material for LIB cathodes is $LiCoO_2$ (LCO). The morphological structure of LCO is a two-dimensional, so-called layered rock structure. It is more stable and has a higher cell voltage and specific capacity than LFP. Although some other two-dimensional materials, such as $LiNiO_2$, LNO, or the three-dimensional $LiMnO_2$ (LMO), do have enhanced properties, they are not practicable as LIB cathodes in their simple form. The three-dimensional, or so-called spinel structure of LMO offers high ion mobility, good stability, and low costs, but it also exhibits a low specific capacity. To combine all the advantages of these various materials, some complex oxide blends, such as $LiNi_{1/3}Mn_{1/3}Co_{1/3}O_2$ (NMC), have been developed. In 2010, NMC was the second most applied cathode material after LCO, with a market share of 40 % (Pistoia, 2014). What all cathode materials have in common is their relatively low electric conductivity. Hence, the particles are coated with conducting additives before being applied as an electrode film (Pistoia, 2014).

Nowadays, most anode materials are variants of graphite, such as artificial, modified natural, or mesophasic graphite (Nazri and Pistoia, 2009; Yoshio et al., 2009; Liu et al., 2010; Gulbinska, 2014; Pistoia, 2014). This material intercalates the lithium-ions into its hexagonal structure with a very low volume expansion (< 10 %) to LiC_6. Other materials with even higher specific capacities, such as $Li_{21}Si_5$, expand in the range of 100 % This significantly reduces the cycle durability of the subsequently-applied anode layers. As a more complex property, graphite has a very low cell voltage against lithium. On the one hand, this allows for the desired high cell voltages in combination with available cathode materials. On the other hand, this low potential enables the deposition of metallic lithium during cell operation. If these so-called dendrites grow until they reach the cathode, they create a short circuit, thus destroying the cell. The anode materials applied in this work were artificial and mostly spherical grahite particles.

8.3 Manufacturing lithium-ion battery electrodes

In the past decades, much experience with lithium-ion batteries has already been gained by using them for small-scale mobile applications, such as mobile phones, notebooks and digital cameras (Consultants, Roland Berger Strategy, 2011a). To date, the cost intensive manufacturing process of suitable large-scale LIB cells has prevented their extensive use in other fields, such as electric mobility or stationary power supply, as described above (Daniel, 2008; Hau, 2008; Consultants, Roland Berger Strategy, 2011a). Besides enhanced cell performance, increased speeds in the production lines and reduced scrap rates could decrease these costs. Here, LIB electrode production offers a high potential for improvements (Daniel, 2008; Consultants, Roland Berger Strategy, 2011b; McKinsey &

Company, 2012; Ruprecht and Kaiser, 2013). Figure 8.3 shows schematically the main process steps for stacked-cell electrode production.

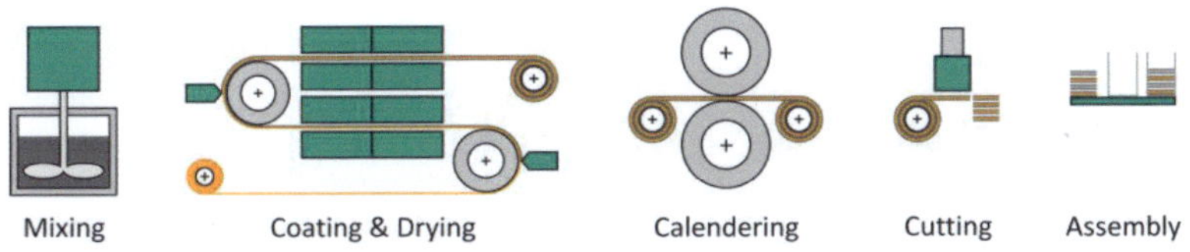

Figure 8.3: Schematic drawing of a typical process chain for stacked LIB electrodes.

The typical LIB electrode film consists of active material particles, binder, and additives (Daniel, 2008; Nazri and Pistoia, 2009; Yoshio et al., 2009; Kaiser et al., 2014). The electrode must be porous to allow for total penetration of the electrolyte as a lithium-ion transport path to the active materials. To deliver current, the electrode needs to be conductive and also connected to suitable metal foils, which serve as current collectors. For the film formation, the electrode components are mixed with a solvent and roll-to-roll coated on the metal substrate as a liquid film. Here, the most precise and common coating method is pre-metered slot die coating (Yoshio et al., 2009). In a subsequent drying step, the solvent is evaporated, leaving the desired porous film structure. To increase adhesion and energy density, the electrodes are calendered to a desired porosity (Nazri and Pistoia, 2009; Yoshio et al., 2009; Zheng et al., 2012). For a stacked cell layout, single stackable electrode sheets are then stamped out of the electrode rolls (Yoshio et al., 2009).

In the following chapters, the challanges, requirements, and potentials of the individual process steps will be discussed.

8.3.1 Powder and slurry mixing

The starring resources in the manufacutring chain of LIB electrodes are the particular existing active materials. The particles vary in size from < 1 µm to 20 µm (Kim et al., 1999; Terashita and Miyanami, 2002; Dominko et al., 2003a; Dominko et al., 2003b). Since the subsequent film coating requires the active materials to be in a liquid phase, the particles are mixed with the needed additives and dispersed in a solvent. Depending on the kind of electrode, cathode or anode, these steps have different requirements (Kaiser et al., 2014).

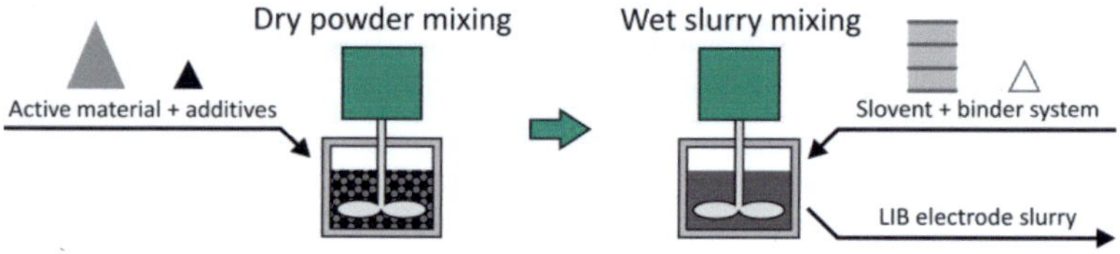

Figure 8.4: The production of LIB electrode slurries is subdivided into powder mixing and slurry dispersion (Kim et al., 1999; Terashita and Miyanami, 2002).

Most cathode active materials exhibit low intrinsic electrical conductivity (Kim et al., 1999; Terashita and Miyanami, 2002; Dominko et al., 2003a; Dominko et al., 2003b; Moskon et al., 2007). However, a minimum conductivity within the electrode films is needed to connect every particle with the current collector. Hence, as plotted in Figure 8.4, in a first step, the cathode materials are dry mixed with a conductive additive, mostly carbon black on the nanoscale. This dry mixing step changes the particle properties (Habermann, 2011) by performing three operations. The carbon black is disagglomerated, milled, and then coated onto the active material particle surface (Dominko et al., 2003b; Dominko et al., 2003a; Moskon et al., 2007; Wenzel et al., 2014; Bitsch et al., 2015). In a second step, a polymeric binder for cohesion and adhesion of the particles to the current collector foil is added. Located between the active particles, this binder allows the finished film to expand and shrink during intercalation. Due to their solubility, polymeric binders are always appropriated by the selected solvent of the subsequent slurry. This might be a combination of polyvinylidene difluoride (PVDF) as binder and n-methyl-2-pyrrolidone (NMP) as solvent (Kim et al., 1999; Yoshio et al., 2009; Neumann, 2011). Either the PVDF is added as a powder in the dry mixing step, or it is added soluted in NMP in the wet mixing step (Kim et al., 1999; Terashita and Miyanami, 2002). During this following wet mixing, all ingredients should be distributed homogeneously. This also includes the remaining free carbon black particles. When embedded in the binder, these free particles ensure an inter-particular conductivity network in the finished film (Moskon et al., 2007; Kaiser et al., 2014).

As the predominant anode active material (Nazri and Pistoia, 2009; Yoshio et al., 2009; Liu et al., 2010; Gulbinska, 2014; Pistoia, 2014), graphite has a relatively high intrinsic conductivity. Therefore, additives to enhance particle surface conductivity are not necessary. Nevertheless, smaller graphites help to form more particle-particle contacts and thus, increase the overall film conductivity. Unlike cathode slurries, state of the art LIB anode slurries are environment-friendly and water-based. Here, the binder system is usually a combination of carboxymethyl cellulose (CMC) and styrene-butadiene rubber (SBR) (Yoshio et al., 2009). Due to its shear-sensitivity, SBR is added at the very end of the wet mixing step, while CMC usually comes pre-dissolved with the water (Kaiser et al., 2014).

As the product of the first electrode manufacturing step, electrode slurries must fullfil certain requirements. First, the cathode particles have to be coated sufficiently with conducting additives. This is crucial for subsequent cell performance and durability (Dominko et al., 2003a; Dominko et al., 2003b; Moskon et al., 2007). Second, all ingredients have to be distributed homogeneously, which excludes any particle agglomerates. Any cluster formation in the future film will restrict its electrical or mechanical performance. Agglomerates on the scale of several active particles might even affect the subsequent coating step (Kaiser et al., 2014; Bitsch et al., 2015). Third, the electrode slurry might need a stable gel character to be storeable. This gel character is a function of the amount of polymeric binder in the slurry and results in visco-elastic behavior (Bitsch et al., 2014). Further, the combination of all slurry ingredients leads to a strong shear-thinning viscosity with relatively high zero shear viscosities (Bitsch et al., 2014; Kaiser et al., 2014; Bitsch et al., 2015). Fourth, all these properties must be reproducible and adjusted to the following coating step to ensure a high quality coating (Gutoff et al., 2006; Kaiser et al., 2014).

8.3.2 Electrode film coating

Within the LIB electrode coating step, the electrode slurries are deposited as a thin liquid film onto the current collector foil. To ensure a high quality coating, there are three crucial prior operations: the slurries must be pumped precisely, they must be pre-treated to increase their homogeneity, and they must be tempered (see Figure 8.5).

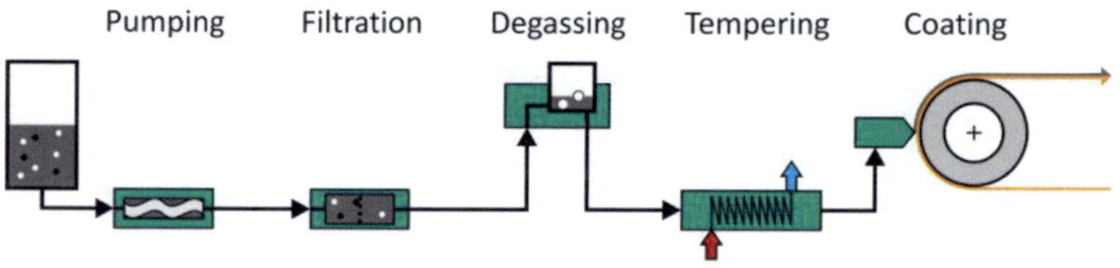

Figure 8.5: For high quality films, the electrode coating requires three prior slurry treatment steps (Cohen and Gutoff, 1992; Gutoff et al., 2006).

Especially for pre-metered coating methods an accurate slurry metering system is crucial (Gutoff et al., 2006). This metering system could work via the principles of gravity, over-pressure, or a mechanical pump (see Figure 8.6). In a system where the slurry feed is provided via gravity, a head tank is installed above the coating equipment. This tank is itself fed by a pump, and an overflow regulates the liquid level. This system requires little maintenance, provides a homogeneous, but low, over-pressure with an unfavorably large liquid hold-up. In contrast, in an over-pressure metering system, the liquid feed is forced out of a tank via a pressurized gas. This system also provides a pulsation-free feed flow with little maintenance, but the continuous refilling of the tank is quite difficult. Finally, a mechanical pumping system offers flexible hold up and quick reaction times, and it is easy to control. The needed maintenance is accomplished without downtime by means of redundancy, while any induced pulsations are dampened by elastic downstream pipes. For battery slurries, the included large and hard particles and the shear-sensitive polymers restrict the variety of suitable mechanical pumps (Cohen and Gutoff, 1992; Kaiser et al., 2014). These could be progressive cavity pumps or, with some limitations to the overall liquid volume, high precision syringe pumps (Gutoff et al., 2006).

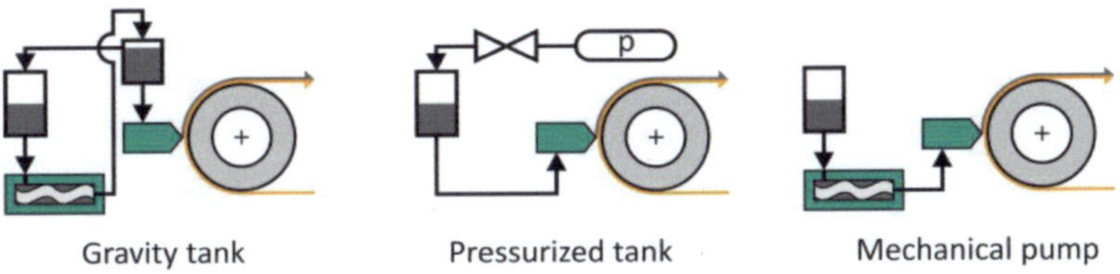

Figure 8.6: Supplying the coating liquid can be accomplished by gravity, a pressurized tank, or a mechanical pump (Cohen and Gutoff, 1992).

The pre-treatment of the slurry incorporates a filtration and a degassing step. Agglomerates or other solid contaminants, such as fibers, affect the ultimate film uniformity and may even provoke a film break-up (Schmitt et al., 2013a) (see Figure 8.7; bottom). Agglomerates might result from an incomplete mixing step or poor colloidal stability. Contaminants may be contained in the liquid almost any time. Dust and packaging residues can be drawn in with the raw materials or may gain entry while the liquid is in contact with the environment. Even in a closed feed system, gasket or oxidation residues can contaminate the liquid. Due to the high LIB slurry viscosities (Bitsch et al., 2014; Kaiser et al., 2014), the application of a metallic mesh filter seems a reasonable solution (Gutoff et al., 2006). Obviously, the mesh must be on the scale of the contaminating particles. Filtration systems having a filter cake formation will increase the pressure drop to an impractical degree.

In addition, gas bubbles can be introduced into the liquid during the mixing step, by a decanting of the liquid, or by air pockets in the piping system (Gutoff et al., 2006). Inside the liquid flow, these bubbles not only cause pressure and thus film fluctuations in machine direction, they may also displace the deposited liquid and result in uncoated areas on the substrate (see Figure 8.7; top-right). Particularly with high-viscous mixing sequences, air is stirred into the battery slurry quite easily (see Figure 8.7; top-left). One way to remove these trapped bubbles and avoid the creation of more is to build in a thin-film-drainage step for every subsequent decanting (Cohen and Gutoff, 1992). To prevent air pockets in the piping, one can avoid high points or install bleed valves in the appropriate places (Gutoff et al., 2006).

The liquid viscosity is strongly temperature-dependent (Wengeler et al., 2013) and, as a major coating parameter, must be known exactly. Therefore, one further important pre-treatment is the pre-heating or cooling of the coating liquid. If the coating apparatus is misused as a final heat exchanger, film-inaccuracies in the cross-machine direction will occur (Cohen and Gutoff, 1992).

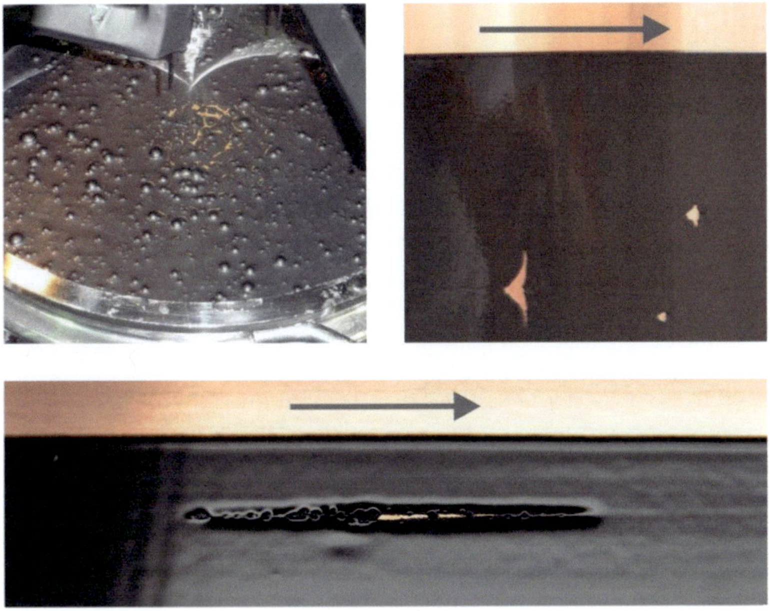

Figure 8.7: Bubbles observed at the surface of a LIB electrode slurry after the mixing step (top-left) and their consequences as uncoated areas in a wet electrode film (top-right). Pinned agglomerates in the coating apparatus also cause uncoated stripes in the wet film until they get flushed out (bottom).

When the electrode slurry is fully homogenized, filtered, bled and tempered, it is suitable for the roll-to-roll coating step. This manufacturing step prearranges the outer shape of the cast film, which has to fulfill multiple requirements: an accurate lateral film thickness, sharp edges without super-elevations, intermittent films, and in certain cases, subdivided layers.

The most crucial requirement is the lateral homogeneity in electrode film thickness (Ketterer B., U. Karl, D. Möst, S. Ulrich, 2009; Fraunhofer-Institut für System- und Innovationsforschung ISI, 2013). Each deviation might lead to local over-charging or more rapid aging in the final LIB cell. On the one hand, the accuracy in machine direction is provided by a precise web-handling and a precise pumping system in the case of a pre-metered coating. On the other hand, the cross-machine accuracy is dependent on the applied coating method. A second requirement concerns the lateral coating edges, which should be perpendicular to the top and bottom of the film in the ideal case (Fraunhofer-Institut für System- und Innovationsforschung ISI, 2013). Besides the above-mentioned consequences for the LIB cell, super-elevations at the edge make it harder to achieve a well-defined roll-to-roll up-winding (Schmitt et al., 2013c). Moreover, the downstream calendering process would also be hindered, as discussed in appendix 8.3.4. Depending on the projected LIB cell layout, a third requirement is a precisely intermitted electrode film (Ozawa, 2009). In relation to the calendering, a periodically interrupted film allows for higher line forces and thus higher energy densities (Zheng et al., 2012). Furthermore, novel cell concepts include subdivided electrodes, in which different layers are stacked to one film (Imachi et al., 2007; Chen et al., 2010; Tran, 2013). In addition to the above requirements, this approach also requires that the different layers are not mixed within the coating.

In the past, self-metering methods, such as the reverse-roll (Ketterer B., U. Karl, D. Möst, S. Ulrich, 2009) and comma-bar coating (Broom, 2011), were applied in electrode production lines. These methods are easy to handle and not very expensive, since they do not require the high precision metering-system described here (Cohen and Gutoff, 1992; Kistler and Schweizer, 1997). Nevertheless, these methods must be calibrated to any environmental change, which impacts the availability of the whole line as well as their precision (Gutoff et al., 2006). Additionally, new requirements, such as subdivided electrode films, are most economical if all layers can be provided simultaneously. This is not practicable with any self-metered coating method.

In today's LIB industry, pre-metered slot die coating is the prevailing coating method (Yoshio et al., 2009; Benhamou, 2011; Glawe, 2014; Mecklenburg, 2014; Briec Emma, Müller Beate, 2015). With its high precision, controllability, reliability, and the means of coating several layers simultaneously, this method fulfills all requirements (Cohen and Gutoff, 1992; Kistler and Schweizer, 1997). The easy scalability of slot die coating in speed and width offers further economic advantages. Usually a slot die coating is applied in the so-called bead mode with a small gap to the electrode foil, which is supported by a backup roller. The foils are as thin as possible, with only 20 µm (aluminium) for the cathode and 10 µm (copper) for the anode. A photo of lab-scale slot die coating of LIB anode films is shown in Figure 8.8. Since this work claims to improve this process, all of its theories, limits, and perspectives will be discussed in more detail in chapter 1.2.

Figure 8.8: Continuous slot die coating of a LIB anode film by the author on the lab scale.

The amount of cast electrode film is determined by the needs of the future LIB cell. As mentioned in appendix 8.2, high power applications require small films with short ionic and electric paths. Conversely, high energy applications need much thicker films for an enhanced specific capacity. Usually this results in dry film thicknesses in the range of 30-200 µm (Ketterer B., U. Karl, D. Möst, S. Ulrich, 2009; Gulbinska, 2014).

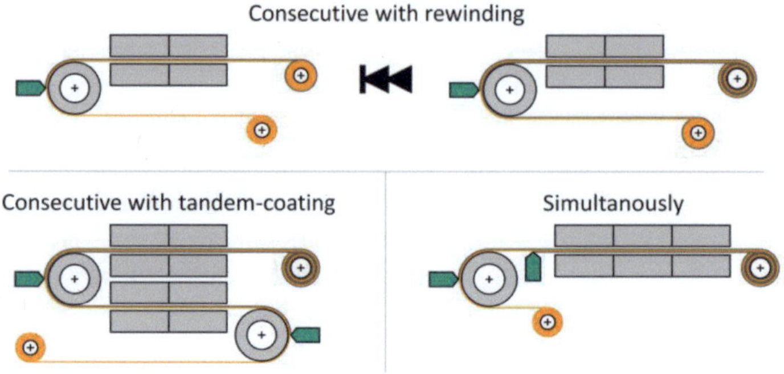

Figure 8.9: The electrode must be coated on both sides. This could accomplished consecutively via rewinding (top) or via a tandem arrangement (bottom, left), or simultaneously via a two-sided system (bottom, right) (Benhamou, 2011; Glawe, 2014; Mecklenburg, 2014; Briec Emma, Müller Beate, 2015).

Due to the multiple-cell battery layouts (see chapter 8.2.1), the current-collector foils are coated on both sides. This can be done in three ways as drawn schematically in Figure 8.9. The most basic, but highly time and labor consuming way is a consecutive procedure with an intervening roll rewinding. Here, the top-side of the foil is first coated and dried before it is rewound and the bottom-side electrode is applied (Mecklenburg, 2014). Note that it might be a disadvantage for the top-electrode morphology to experience the dryer-heat a second time. A less time-consuming, but more investment intensive way is the consecutive procedure on a tandem coater. By integrating a second coating and drying step into the production line, the rewinding is made unecessary (Benhamou, 2011). Here again, however, the first applied electrode experiences more heat and the increased investment costs require a high throughput for an economical production. The most advanced double-sided electrode production is realized by a simultaneous coating procedure. Here, the electrode foil is coated on both sides, before it is dried in a single dryer (Briec Emma, Müller Beate, 2015). The double-sided coating step is a challenge which again can only be met using pre-metered coating methods. Slot die coating for example, does not necessarily require a backing roller for operation. It can also be applied in a so-called tensioned-web mode (Kistler and Schweizer, 1997). Thus, several arrangements are possible, for example a combination of bead mode on the top side and tensioned-web mode on the bottom side. Or, both electrode sides could be coated in the tensioned-web mode. The single dryer then needs air floatation nozzles to support the double-sided wet foil and has to evaporate the increased quantity of solvent.

8.3.3 Electrode film drying

While the outer shape of the electrode film forms during the coating, the height shrinks and the internal morphology develops during the solvent-evaporating drying step (Kaiser et al., 2014; Baunach et al., 2015). This development includes the distribution and arrangement of all ingredients, as well as the film porosity during film shrinkage. The reqirements for the film morphology are to provide sufficient adhesion and cohesion and good ion and electric conductivity (Ketterer B., U. Karl, D. Möst, S. Ulrich, 2009; Neumann, 2011; Fraunhofer-Institut für System- und Innovationsforschung ISI, 2013).

In LIB electrode production lines, the drying step directly follows the film coating step. To enhance energy absorption, the applied convective impingement-jet dryers can be supplemented by infrared heaters. As mentioned above, simultaneous, double-side coated electrodes require an air floatation system capable of carrying the wet foil (Benhamou, 2011; Glawe, 2014). For LIB electrode films, the residence time in the dryer is 1-2 minutes. For today's production speeds of 50 m/min, this results in dryer lengths on the order of ≤ 100 m.

A dryer is traditionally divided into several zones having different conditions. It has been observed that the distribution of additives and binder affect the cell aging process (Vetter et al., 2005; Barré et al., 2013). However, the homogeneous distribution of the additive particles and polymeric binders in the electrode slurry is reversed during solvent evaporation (Haselrieder et al., 2013; Kaiser et al., 2014). The still unknown, prevailing kinetics and their trends are the subjects of various scientific works (Hagiwara, 2012; Baunach, 2014; Jaiser, 2015). It was found that drying rate and temperature affect the film adhesion force, which is an indicator for the binder arrangement (Baunach et al., 2015). Based on empirical evidence, when the dryer zones of industrial production lines are adjusted to a low-high drying mode, it leads to best performance (Mecklenburg, 2014). Here, the first dryer zones apply low temperatures and drying rates and these are followed by zones with higher temperatures and drying rates. The most recent investigations have decoupled temperature and drying rate, which is also a function of the mass transfer coefficient and humidity (Schlünder, 1996). It was demonstrated that the zone adjustment could be improved, thus leading to shorter residence times and shorter dryers (Baunach et al., 2015; Jaiser, 2015).

8.3.4 Electrode calendering

The solidified electrode film leaves the drying step with a relatively high porosity of approximately 60 % (Baunach et al., 2015). To increase the energy density, the films are calendered, decreasing the film height and thus the porosity to approximately 30-35 % (Kaiser et al., 2014). It was shown that compacting the film structure also increases its adhesion to the current collector foil (Zheng et al., 2012; Baunach et al., 2015). The resistivity also decreases for more compact morphologies (Terashita and Miyanami, 2002; Haselrieder et al., 2013). However, below an optimum porosity, the ion-conductivity and thus the usable capacity decreases (Zheng et al., 2012).

The requirements here are quite similar to those for the previous coating step. The deviations in the resulting film height and thus the porosity must be kept very small to avoid local current power peaks or power resistance spots (Kaiser et al., 2014). As a consequence, inhomogeneities arising during the coating step must also be avoided.

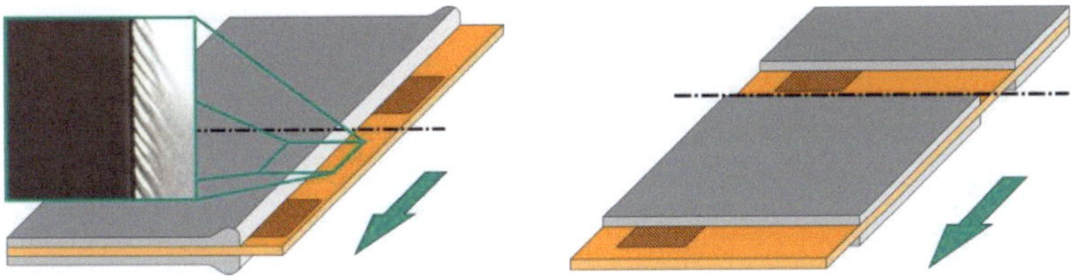

Figure 8.10: Calendered, continuous coating with super-elevated edges and wrinkles in the uncoated area (left); intermittent coating without uncoated areas (right) (Schmitt et al., 2015a).

During the calendering step, the film and the coated part of the metal foil are stretched. Depending on the calendering pressure, this stretching can lead to undesired folds at the intersections between coated and uncoated areas of the substrate (Figure 8.10, left). Amplified by super-elevations at the edge of the film, these folds can cause film-peeling and prevent precise electrode cutting in the cell assembly step. One way to prevent this critical defect is to cut off the uncoated substrate and the edge of the coating. Due to the need for an uncoated current collector, this is not possible for continuously coated substrates. For this reason, some manufacturers coat intermittently and place the current collectors in the uncoated gaps between the film patterns (see Figure 8.10, right) (Mecklenburg, 2014). This enables one to cut off the critical edge region, thus preventing problems during calendering.

8.3.5 Electrode cutting

The overall length of roll-to-roll processed raw-electrodes can be as long as 2000 m at widths of 1000 mm (Kaiser et al., 2014). This is far larger than the dimensions of the future LIB cells (Yoshio et al., 2009). Hence, the rolls must be slit lengthwise and, depending on the cell layout, single electrode sheets have to be die cut (Kurfer et al., 2012; Baumeister and Fleischer, 2014).

The requirements for these processes are diverse. Due to the very abrasive cathode materials, their slitting is performed by ceramic blades. Even micrometer-scale metallic swarfs could damage the sensitive separator and provoke short circuits in the finished cell (Kaiser et al., 2014). Furthermore, the handling of the die-cut of very limp electrode sheets destined for stacked cells requires a large procedural effort (Fleischer et al., 2012; Ebert et al., 2014). New developments reduce this effort by combining the advantages of stacked cell layouts with a folding assembly process (KIT, 2014).

At this point, the LIB electrodes are ready for the cell assembling process, which includes stacking, packaging, electrolyte filling and sealing, followed by the first electric charging (Yoshio et al., 2009; Kaiser et al., 2014).

8.4 Redistribution of sub-µm particles in simultaneous bilayer coatings

To investigate the possibilities of bilayer coatings, LIB anodes were produced on a pilot-scale roll-to-roll coater (see Figure 8.11).

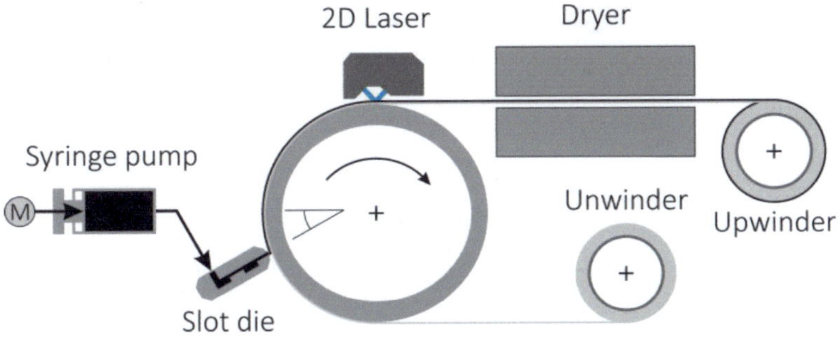

Figure 8.11: Schematic diagram of the pilot-scale roll-to-roll plant for the manufacture of LIB electrodes.

It was planned to improve the energy density of an anode by reducing the additive content of its upper part. It was postulated that the majority of binders and conductive additives are needed close to the conducting foil. The improved anode was realized by means of simultanous bilayer coating with slot die 2 and $m = 2$, where slurry 2 was coated as the bottom layer. In addition, an improved slurry 2, with a lowered binder content and no conductive additives, slurry 2*, was coated as the top layer.

To evaluate the resulting bilayer anode, single-layer anodes of slurry 2, improved slurry 2*, and a mixture of both adjusted to the feed ratio were coated as well, to allow for comparison. It was expected that the adhesion force of the bilayer film and slurry 2 would be equal, since their binder content at the substrate is the same. Analogously, the adhesion force of the films containing the mixture of both slurries and plain slurry 2* was expected to be lower. The results are plotted in Figure 8.12

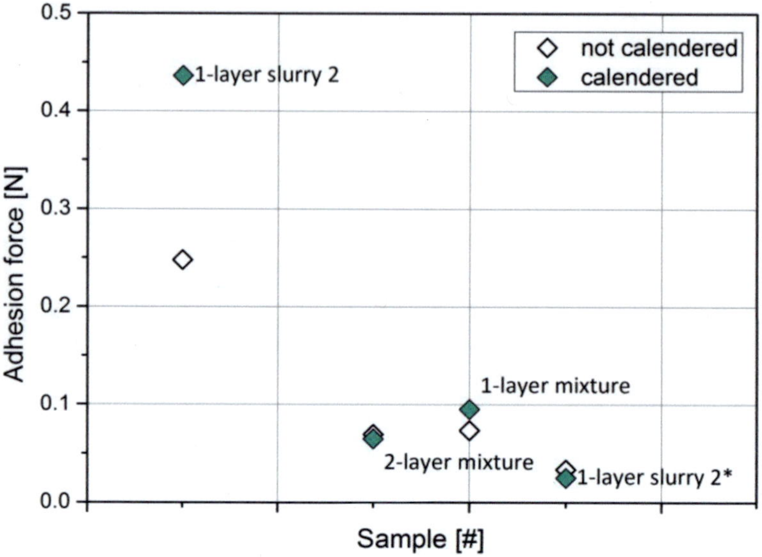

Figure 8.12: Resulting adhesion forces of the different anodes before and after calendering.

As the above chart shows, the adhesion force of the bilayer film is at the same level as the anode mixture. This might be explained by a rearrangement of the pre-distributed binder during the drying step. Focusing on the non-calendered films, the levels of adhesion force of plain slurry 2*, the mixture of both, and plain slurry 2 roughly reflect the total amount of included binder. One also sees that the adhesion force of the low level samples cannot be improved appreciably by the calendering step. Even though these trials show that pre-distribution of the binder between top and bottom layers might be revised during drying, the literature show that this effect can be controlled (Baunach et al., 2015; Jaiser, 2015).

8.5 The pendant drop method

A liquid's surface tension can be calculated on the basis of the measured drop dimensions (see Figure 8.13) during the microscopic image acquisition of the pendant drop method. The required dimensions are the diameters d_1 and d_2, the vertical drop length l_{Drop}, the drop density φ_{Drop}, and the gravity g:

$$\sigma = \frac{\varphi_{\mathrm{Drop}} \cdot g \cdot l_{Drop}}{2 \cdot \left(\frac{1}{d_1} + \frac{1}{d_2}\right)} \tag{8.5}$$

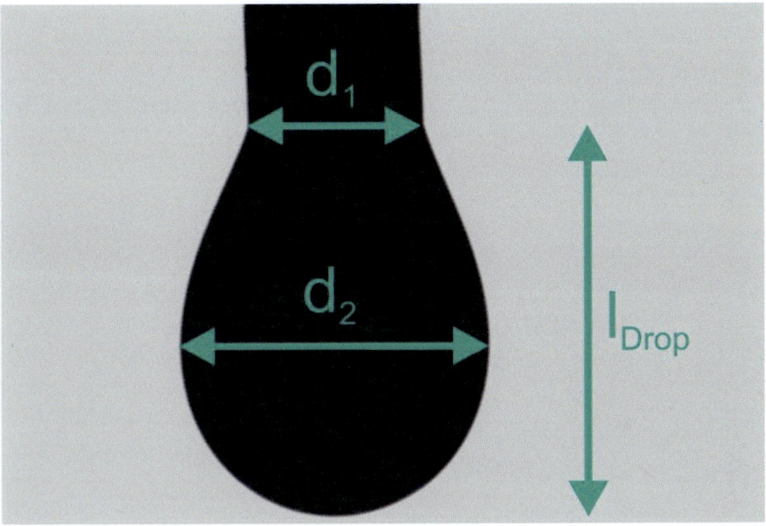

Figure 8.13: Required dimensions for calculating surface tension using the pendant drop method.

Nevertheless, this method is very subjective. It deviates between different operators and can also be also affected by vibrations and solvent drying (KRÜSS GmbH; Menges, 2002).

8.6 Numerical simulation of bead stability

In a very basic numerical simulation with ~1,000,000 cells, the coating gap region of slot die 1 with shear-thinning slurry 1 at $u_W = 85\ m/min$ was computed unsteady.

As shown in Figure 8.14 (top), the upstream meniscus wets the upstream gap section partially for $h = 84\ \mu m$, while the coating bead is stable. In contrast, the upstream meniscus runs back to the feed slot for $h = 54\ \mu m$. It is over-stretched and thus air is entrained in the coating bead, which results in periodic uncoated parts in the cast film.

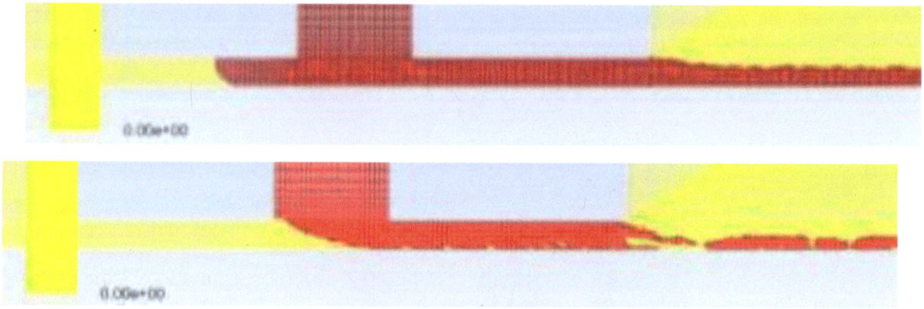

Figure 8.14: Screenshots of the numerically simulated gap flow of slot die 1 with a film thickness of $h = 84\ \mu m$ (top) and $h = 53\ \mu m$ (bottom) at $u_W = 85\ m/min$.

8.7 Microscopic image acquisition

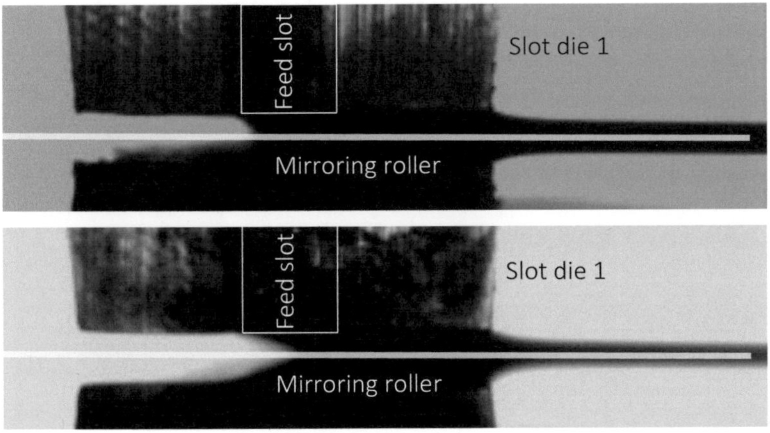

Figure 8.15: Microscopic image acquisition of the gap region of slot die 1 during coating of slurry 1 at $u_W = 4\,m/min$. The film thickness was decreased from $h = 90\ \mu m$ (top) to $39\ \mu m$ (bottom).

The above images show that the upstream meniscus of the coating bead is pinned at its final position at the corner of the feed slot for $u_W = 4\,m/min$. By decreasing the film thickness from $h = 90\ \mu m$ (top) to $39\ \mu m$ (bottom), the meniscus then becomes overstreched and air entrainment occurs periodically. In contrast, the downstream meniscus is pinned steadily at the downstream corner of the die.

The pictures of the slot die 2 coating gap under similar conditions, presented in Figure 8.16, confirm these results for bilayer applications.

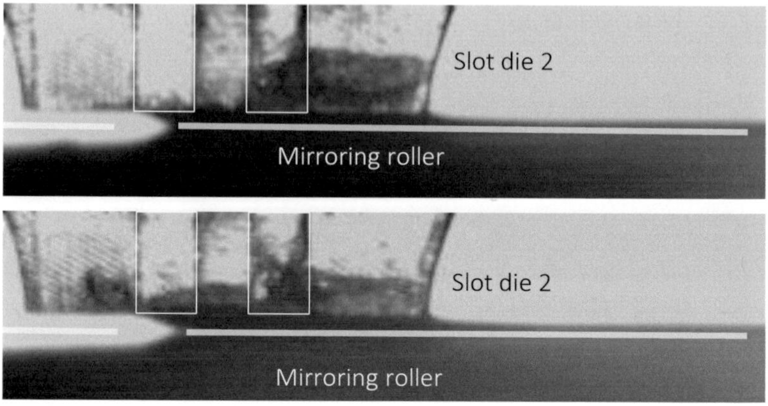

Figure 8.16: *Microscopic image acquisition of the gap region of slot die 2 during bilayer coating of slurry 1 at* $u_W = 4\ m/min$*.*
The film thickness was decreased from $h = 70\ \mu m$ *(top) to* $52\ \mu m$ *(bottom).*

8.8 Dynamic upstream capillary pressure

In addition to the simplified expression (see equation (3.7)), the upstream capillary pressure drop can be described more precisely with regard to the wettability of the die lips and the moving web (Higgins and Scriven, 1980):

$$\Delta p_{3U} = \Delta p'_{4U} = \frac{\sigma}{h_G}(\cos\vartheta + \cos\theta) \tag{8.6}$$

Here, $\Theta = 25.9°$ represents the static contact angle of slurry 1 on steel, whereas ϑ stands for the dynamic contact angle, which can be calculated as follows (Cohen and Gutoff, 1992; Kistler and Schweizer, 1997):

$$\vartheta = 73 \cdot u_W^{0.22} \cdot \eta^{0.18} \cdot \sigma^{0.11} \tag{8.7}$$

However, if equation (8.6) is substituted for equation (3.7) in the calculation for the minimum film thickness, the resulting differences are too small to change the overall result. In Figure 8.17, the comparison of the computed results with a static and a dynamic upstream capillary pressure drop are plotted.

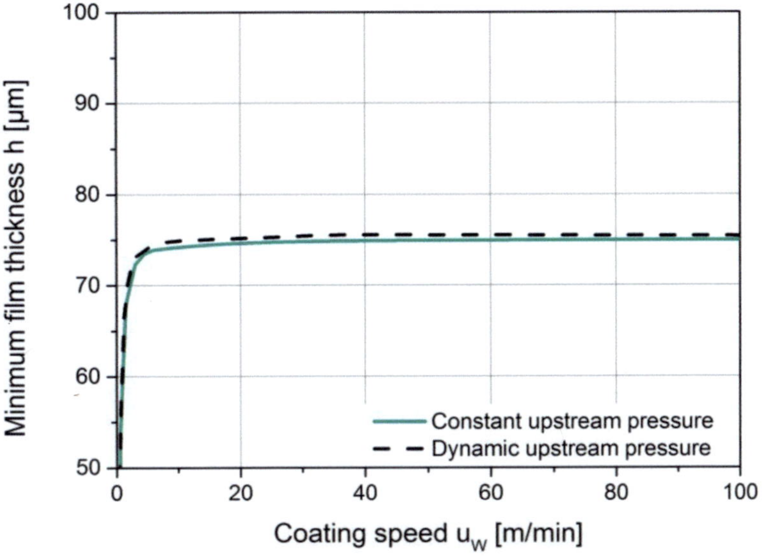

Figure 8.17: *Comparison of the computed minimum film thickness of slurry 1 with slot die 1 at* $h_G = 148\,\mu m$ *for a static and a dynamic upstream capillary pressure drop.*

8.9 Dampening at intermittent coatings

The experiments on long-time dampening in intermittent systems with too high waste-pipe pressures (presented in Figure 5.5) were repeated for intermittent coatings with too low waste pipe pressures. These results are plotted in Figure 8.18.

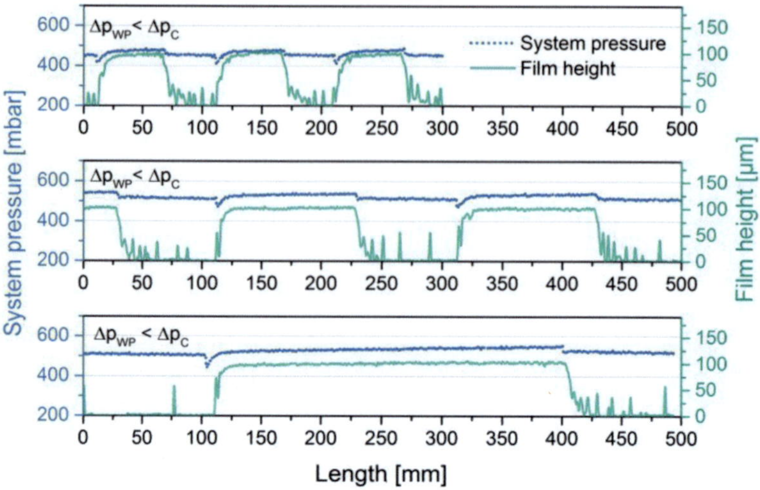

Figure 8.18: *System pressure and the related film thickness for intermittent coatings at $u_W = 1\, m/min$ and a too low waste pipe pressure drop plotted for three different period lengths. The lengths were varied from $60/40\, mm$, to $120/80\, mm$ and $300/200\, mm$ for the coating and interruption periods, respectively.*

Similar to the trend for too high waste pipe pressures, the trend for too low waste pipe pressures is only observable for longer coatings (Figure 5.5, bottom). Even after $l_C = 300\, mm$ of coating, the preset film height was not reached.

List of publications

Publications

[1] **M. Schmitt**, S. Raupp, D. Wagner, P. Scharfer, W. Schabel. *Analytical determination of process windows for bilayer slot die coating.* Journal of Coatings Technology and Research, Volume 12, Issue 5, Page 877-887, August (2015).

[2] **M. Schmitt**, R. Diehm, P. Scharfer, W. Schabel. *An experimental and analytical study on intermittent slot die coating of viscoelastic battery slurries.* Journal of Coatings Technology and Research, Volume 12, Issue 5, Page 927-938, August (2015).

[3] J. Kaiser, V. Wenzel, H. Nirschl, B. Bitsch, N. Willenbacher, M. Baunach, **M. Schmitt**, S. Jaiser, P. Scharfer, W. Schabel. *Prozess- und Produktentwicklung von Elektroden für Lithium-ionenzellen.* Chemie Ingenieur Technik, Volume 86, Issue 5, Pages 695–706, May, (2014).

[4] B. Bitsch, J. Dittmann, **M. Schmitt**, P. Scharfer, W. Schabel, N. Willenbacher. *A novel slurry concept for the fabrication of Lithium-Ion battery electrodes with beneficial properties.* Journal of Power Sources, Volume 265, 1 November 2014, Pages 81–90

[5] L. Wengeler, K. Peters, **M. Schmitt**, T. Wenz, P. Scharfer, W. Schabel. *Fluid-dynamic properties and wetting behavior of coating inks for roll-to-roll production of polymer-based solar cells.* Journal of Coatings Technology and Research, January 2014, Volume 11, Issue 1, pp 65-73.

[6] **M. Schmitt**, P. Scharfer, W. Schabel.
Slot die coating of lithium-ion battery electrodes: investigations on edge effect issues for stripe and pattern coatings. Journal of Coatings Technology and Research, January 2014, Volume 11, Issue 1, pp 57-63

[7] **M. Schmitt**, M. Baunach, L. Wengeler, K. Peters, P. Junges, P. Scharfer, and W. Schabel.
Slot-die processing of lithium-ion battery electrodes–Coating window characterization. Chemical Engineering and Processing: Process Intensification, Volume 68, June 2013, Pages 32–37.

[8] L. Wengeler, **M. Schmitt**, K. Peters, P. Scharfer, and W. Schabel.
*Comparison of large scale coating techniques for organic and hybrid films in polymer based solar cells.*Chemical Engineering and Processing: Process Intensification, Volume 68, June 2013, Pages 38–44

[9] L. Wengeler, **M. Schmitt**, P. Scharfer, W. Schabel.
Designing a sensor for local heat transfer in impingement driers. Chemical Engineering and Processing: Process Intensification, 50, no. 5 (2011): 516-518.

Conference contributions

[1] **M. Schmitt**, P. Kitz, P. Scharfer, W. Schabel
Edge formation in slot die coating of lithium-ion battery electrodes (Talk). European Coating Symposium 2015 (ECS15), 09.-11.09.2015, Eindhoven, Netherlands

[10] **M. Schmitt**, M. Baunach, S. Jaiser, P. Scharfer, W. Schabel
Coating and Drying of Lithium-Ion Battery Electrodes (Poster). Materials Science Engineering (MSE 2014),23.-25.09.2014, Darmstadt, Germany

[11] **M. Schmitt**, P. Scharfer, W. Schabel
Multilayer slot die coating of Li-ion battery electrodes (Talk). 17th International Coating Science and Technology (ISCST 2014) Symposium, 7.-10.09.2014, San Diego CA, USA

[12] **M. Schmitt**, R. Diehm, P. Scharfer, W. Schabel
New intermitting concepts for slot die coating of Li-ion battery electrodes (Talk). 17th International Coating Science and Technology (ISCST 2014) Symposium, 7.-10.09.2014, San Diego CA, USA

[13] **M. Schmitt**, B. Bitsch, N. Willenbacher, P. Scharfer, W. Schabel
Coating of non-Newtonian lithium-ion battery electrode slurries (Poster). 9th Annual European Rheology Conference (AERC 2014), 08.-11.04.2014, Karlsruhe, Germany

[14] M. Baunach, **M. Schmitt**, S. Jaiser, P. Scharfer, W. Schabel
Batterien für die Versorgungssicherheit der Zukunft – Grundlagenforschung in der Prozessierung von Batterieelektroden (Poster). EnergieCampus 2013, 14. November 2013, Ulm, Germany

[15] **M. Schmitt**, P. Scharfer, W. Schabel
Stripe and pattern coatings of slot die coated battery electrodes – investigation on edge effects (Talk). European Coating Symposium 2013 (ECS13), 10.-13.09.2013, Mons, Belgium

[16] **M. Schmitt**, M. Carvalho, P. Scharfer, W. Schabel
Break-up phenomena in multilayer slot die coating of lithium-ion batteries (Awarded Poster). European Coating Symposium 2013 (ECS13), 10.-13.09.2013, Mons, Belgium

[17] **M. Schmitt**, M. Baunach, P. Scharfer, W. Schabel
Surface quality of slot-die coated battery electrodes with special attention to edge effects (Poster). Materials Science Engineering (MSE 2012),25.-27.09.2012, Darmstadt, Germany

[18] L. Wengeler, **M. Schmitt**, K. Peters, P. Scharfer, W. Schabel
Coating and wetting of semiconducting organic and hybrid films: Fluid-dynamic properties, process parameters, and wetting behavior (Talk). Materials Science Engineering (MSE 2012),25.-27.09.2012, Darmstadt, Germany

[19] M. Baunach, S. Jaiser, **M. Schmitt**, P. Scharfer, W. Schabel
Impact of solidification conditions on film properties of lithium-ion battery electrodes (Poster). Materials Science Engineering (MSE 2012),25.-27.09.2012, Darmstadt, Germany

[20] **M. Schmitt**, M. Baunach, P. Scharfer, W. Schabel
Wet film characterization of slot-die coated multilayer battery electrodes (Talk). 16th International Coating Science and Technology (ISCST 2012) Symposium, 9.-12.09.2012, Atlanta GA, USA

[21] **M. Schmitt**, M. Baunach, P. Scharfer, W. Schabel
Surface quality of slot-die coated battery electrodes with special attention to edge effects (Poster). 16th International Coating Science and Technology (ISCST 2012) Symposium, 9.-12.09.2012, Atlanta GA, USA

[22] L. Wengeler, **M. Schmitt**, K. Peters, P. Scharfer, W. Schabel
Coating and Wetting of Semiconducting Organic and Hybrid Films: Fluid-dynamic Properties, Process Parameters, and Wetting Behavior (Talk). 16th International Coating Science and Technology (ISCST 2012) Symposium, 9.-12.09.2012, Atlanta GA, USA

[23] M. Baunach, **M. Schmitt**, S. Jaiser, P. Scharfer, W. Schabel *Untersuchung des Prozesseinflusses auf die Morphologieausbildung bei Elektroden und Separatoren für Lithium-Ionen Batterien* (Poster). Jahrestagung, KIT-Zentrum für Energie, 19. Mai 2012, Karlsruhe, Germany

[24] **M. Schmitt**, L. Wengeler, P. Darmstädter, C. Kroner, K. Peters, K. Köhler, P. Niyamakom, P. Scharfer and W. Schabel, *Comparison of large scale coating techniques for polymer particle films in hybrid solar cells* (Poster). 8th European Congress of Chemical Engineering, 26-29.09.2011, Berlin, Germany

[25] L. Wengeler, **M. Schmitt**, P. Darmstädter, C. Kroner, K. Peters, K. Köhler, P. Niyamakom, P. Scharfer and W. Schabel, *Comparison of large scale coating techniques for polymer particle films in hybrid solar cells* (Talk). 9th European Coating Symposium, 08-10.06.2011, Abo/Turku, Finland

[26] L. Wengeler, **M. Schmitt**, P. Scharfer, W. Schabel *Entwicklung eins Sensors zur Messung des lokalen Wärmeübergangs in Prallstrahltrocknern* (Poster). ProcessNet 2010, Aachen, Germany

[27] L. Wengeler, **M. Schmitt**, P.Scharfer, W. Schabel, *Designing a sensor for local heat transfer in impingement driers* (Poster). 15th International Coating Science and Technology Symposium, September 12-15, 2010, St. Paul, MN, USA

[28] L. Wengeler, **M. Schmitt**, P. Scharfer, W. Schabel. *Entwicklung eines Sensors zur Messung des lokalen Wärmeübergangs in Prallstrahltrocknern.* Chemie Ingenieur Technik, 82, no. 9 (2010): 1428-1429.

Student theses conducted in conjunction with this thesis

[1] **Manuel Schuster**
Investigations and high speed automation of intermittent slot die coating. Diploma thesis, KIT 2015

[2] **Sandro Spiegel**
Investigations on the system pressure at intermittent lithium-ion battery electrode coatings. Bachelor thesis, KIT 2014

[3] **Ralf Diehm**
Investigations on scalability of intermittent and multilayer coated lithium-ion battery electrodes. Diploma thesis, KIT 2014

[4] **Marco Branghofer**
Untersuchung des Einflusses von Polyvinylidenflourid-Binder auf die rheologischen Eigenschaften von N-Methyl-2-pyrrolidon-basierten Kathodenslurries. Bachelor thesis, KIT 2013

[5] **Laura Rolinger**
Characterization of slot-die coated lithium-ion battery anodes and cathodes with special attention to edge effects. Bachelor thesis, KIT 2013

[6] **Susanne Unger**
Investigations on process conditions for slot die casted lithium-ion battery electrodes. Diploma thesis, KIT 2013

[7] **Paul Kitz**
Characterization of a new slot die design for intermittent lithium-ion battery coatings. Bachelor thesis, KIT 2013

[8] **Moritz Weiß**
Simulation of liquid flow in the coating gap during multilayer slot-die coating of lithium-ion battery electrodes. Bachelor thesis, KIT 2013

[9] **Michael Wittek**
Simulation of liquid flow in the coating gap during slot-die coating of lithium-ion battery electrodes. Bachelor thesis, KIT 2012

[10] **Miriam Vogt**
Investigations on a newly developed slot die concept for intermittent lithium-ion battery electrode coatings. Bachelor thesis, KIT 2012

[11] **Niklas Russner**
Simulation of the interior liquid flow of slot dies during the coating of lithium-ion battery electrodes. Bachelor thesis, KIT 2012

[12] **Dennis Wagner**
Investigation on coating limits of multilayer coated lithium-ion battery electrodes. Bachelor thesis, KIT 2012

[13] **Cornelia Bürkin**
Characterization of slot-die coated lithium-ion battery electrodes with special attention to edge effects. Bachelor thesis, KIT 2012

[14] **Robert Schneider**
Investigation on multilayer coating of lithium-ion battery electrodes. Student research project (Studienarbeit), KIT 2012

[15] **Pascal Junges**
Investigation on the coating behavior of lithium-ion battery electrode slurries. Diploma thesis, KIT 2011

Curriculum vitae

Marcel Schmitt

Personal Data

Date of birth:	March 13th, 1986
Place of birth:	Worms, Germany
Citizenship:	German

Professional Experience

Since 09/2015	Research Engineer BASF Advanced Chemicals Co., Ltd., Shanghai, China
02/2011 – 08/2015	Research assistant / PhD candidate Karlsruhe Institute of Technology (KIT) Institute of Thermal Process Engineering Workgroup „Thin Film Technology“ Supervisor: Prof. Dr.-Ing. Dr. h. c. Wilhelm Schabel

Education

12/2015	**Doctor in Process Engineering** (Slot die coating of lithium-ion battery electrodes), KIT (with distinction)
10/2010	Degree **Process Engineer,** Karlsruhe Institute of Technology (KIT)
10/2005 – 10/2010	**Studies in process engineering,** Karlsruhe Institute of Technology (KIT) Former Research University Karlsruhe (TH)
Major subjects:	Thermal process engineering, Plant engineering
Minor subjects:	Drying technology of thin films, Energy technology
Extra subject:	Micro process engineering
Diploma thesis:	“Commissioning and parameter-optimization of a pilot-plant for the manufacturing of organic and hybrid polymeric solar cells”
04/2005	**Abitur (High school diploma),** Gauß-Gymnasium Worms

Awards

2013	Best Poster Award: “Break-up phenomena in multilayer slot die coating” *European Coating Symposium, September 10th-13th 2013, Mons, Belgium*
2015	KIT “Neuland” Innovation Award: “Novel slot die” *April 23th 2015, Karlsruhe, Germany*

Personal Interests

Sports	Rowing, Surfing, Snowboarding
Music	E-Guitar
Other	Traveling, Design, History